# Making Medical Knowledge

How is medical knowledge made? New methods for research and clinical care have reshaped the practices of medical knowledge production over the last forty years. Consensus conferences, evidence-based medicine, translational medicine, and narrative medicine are among the most prominent new methods. *Making Medical Knowledge* explores their origins and aims, their epistemic strengths, and their epistemic weaknesses. Miriam Solomon argues that the familiar dichotomy between the art and the science of medicine is not adequate for understanding this plurality of methods. The book begins by tracing the development of medical consensus conferences, from their beginning at the United States' National Institutes of Health in 1977, to their widespread adoption in national and international contexts. It discusses consensus conferences as social epistemic institutions designed to embody democracy and achieve objectivity. Evidence-based medicine, which developed next, ranks expert consensus at the bottom of the evidence hierarchy, thus challenging the authority of consensus conferences. Evidence-based medicine has transformed both medical research and clinical medicine in many positive ways, but it has also been accused of creating an intellectual hegemony that has marginalized crucial stages of scientific research, particularly scientific discovery. Translational medicine is understood as a response to the shortfalls of both consensus conferences and evidence-based medicine. Narrative medicine is the most prominent recent development in the medical humanities. Its central claim is that attention to narrative is essential for patient care. Solomon argues that the differences between narrative medicine and the other methods have been exaggerated, and offers a pluralistic account of how the all the methods interact and sometimes conflict. The result is both practical and theoretical suggestions for how to improve medical knowledge and understand medical controversies.

**MIRIAM SOLOMON** is Professor of Philosophy at Temple University. She has a BA in Natural Sciences from Cambridge University and a PhD in Philosophy from Harvard University. Her first book was *Social Empiricism* (MIT Press, 2001) and she is the author of many articles in philosophy of medicine, philosophy of science, gender and science, epistemology, and bioethics. She is a Fellow of the College of Physicians of Philadelphia.

# Making Medical Knowledge

Miriam Solomon

OXFORD
UNIVERSITY PRESS

**OXFORD**
UNIVERSITY PRESS

Great Clarendon Street, Oxford, OX2 6DP,
United Kingdom

Oxford University Press is a department of the University of Oxford.
It furthers the University's objective of excellence in research, scholarship,
and education by publishing worldwide. Oxford is a registered trade mark of
Oxford University Press in the UK and in certain other countries

Published in the United States of America by Oxford University Press
198 Madison Avenue, New York, NY 10016, United States of America

British Library Cataloguing in Publication Data
Data available

Library of Congress Cataloging in Publication Data
Data available

ISBN 978–0–19–873261–7 (Hbk.)
ISBN 978–0–19–287229–6 (Pbk.)

Cover image: © iStock.com/TheStockLab

*To John*

# Preface and Acknowledgments

*Making Medical Knowledge* is a book about the epistemology of medicine. It developed in part from my earlier work on the social epistemology of science (Solomon 2001) and in part from my longstanding interests in the philosophy of medicine. The initial engagement was with the epistemology of medical consensus conferences. When I became aware of them in the 1990s[1] they were already established as a source of authoritative expert statements, but recently challenged by the newer standards of evidence-based medicine, which typically rank expert consensus very low in evidence hierarchies. Moreover, a reaction to evidence-based medicine was being expressed under the banner of the "art" versus the "science" of medicine, perhaps most prominently by those working with the new techniques of "narrative medicine." While I was puzzling over these developments, another new approach—translational medicine—was added to the mix. As I pursued my questions, the outline of a book on new epistemological approaches in medicine took shape, and began in earnest around 2007. It was not until academic year 2012–13, however, that I had the time to write a full draft. I am grateful to Temple University for granting me sabbatical leave, and to the Science, Technology, and Society Program at the National Science Foundation for a Scholar's Award (SES-1152040) supplementing my sabbatical, during 2012–13.

At the same time as I was writing this book, I participated in the steering committee of the new International Philosophy of Medicine Roundtable,[2] which developed from two conferences organized by Harold Kincaid, then at the University of Alabama, Birmingham (in 2005 and 2008). At these and at subsequent Roundtables (2009, 2011, and 2013), I have benefited from presenting my work in progress and from extensive conversations about the epistemology and metaphysics of medicine. It

---

[1] Because of my interests in the epistemology of consensus, I was invited to participate in a consensus conference of the Eastern Association for the Surgery of Trauma in 1995. Before that time I did not know about consensus conferences in medicine, which were begun at NIH in 1977.

[2] <http://philosmed.wordpress.com/> accessed October 30, 2014.

has been exciting to witness and contribute to the development of a new area of study.

The Greater Philadelphia Philosophy Consortium (GPPC), together with the Philadelphia Area Center for the History of Science (PACHS), sponsors a History and Philosophy of Science reading group that meets monthly. Recently we have been reading and discussing in the area of historical epistemology, which is part of the intellectual background for this book. I am grateful to Babak Ashrafi, the Executive Director of PACHS, for hosting the reading group at PACHS, and to Gary Hatfield, who co-organizes the reading group with me.

The Center for the Humanities at Temple University (CHAT), in the College of Liberal Arts, supported my work on medical consensus conferences with a Faculty Fellowship in 2008–9. It also provided a local intellectual community, especially by hosting the Medical Humanities Reading Group from 2010–12. My thanks to Peter Logan, the Director of CHAT, and to Gretchen Condran and Sue Wells, who ran the Medical Humanities Reading Group with me.

My work in progress on this book was presented at a number of academic institutions, to philosophy, science studies, and medical audiences. I learned much from the comments of those in the audience. In particular, I thank the audiences at the Georgia Institute of Technology, The Johns Hopkins University, the University of Washington Seattle, Drexel University, the University of Pennsylvania, the University of Sydney, the University of Miami, the University of South Carolina, Wesleyan University, and Rowan University.

Many individuals—colleagues, students, friends, and family—have supported the writing of this book. Nickolas Pappas and Marya Schechtman responded constructively to the roughest of drafts. Robyn Bluhm read and commented on at least half of the book in progress, and John Ferguson, Federica Russo, and David Teira gave feedback on Chapters 2, 5, and 7 respectively. John Clarke provided on-the-spot medical expertise, and proofread the entire manuscript with the goal of helping to make it accessible to a medical as well as to a philosophical audience. Ellen Peel has been helping me find protected time to write for over 25 years; she is a true work ally. Temple Philosophy librarian Fred Rowland provided invaluable information and advice about literature searches and the organization of references. I also want to thank (in alphabetical order) those who have responded to particular ideas in the book, and/or offered general encouragement and ideas: Kenneth Bond, Havi Carel, Nancy Cartwright, Vivian Coates, Gretchen Condran, Arthur Elstein, Julia Ericksen, John Ferguson, Arthur Fine, Fred Gifford, Kenneth Goodman, Gary Hatfield, Jeremy Howick, Anastasia Hudgins, Harold Kincaid, Fred Kronz, Susan Lindee, Hilde Lindemann, Francesco Marincola, Lenny Moss, Steve Peitzmann, Alan Richardson, Federica Russo, Peter Schwartz, Sandy Schwartz, Judy Segal, Jeremy Simon, Hal Sox, David Teira, Sue Wells, John Worrall, and Alison Wylie. Two anonymous reviewers for Oxford University Press provided detailed and valuable feedback, which I have used in revising the manuscript. I could not have written the book without all this help and support, and I am exceedingly grateful for it.

This book is dedicated, with all my love, to my husband, John R. Clarke. He supported the project at every stage, talked with me about it whenever I asked, and tolerated my occasional need to shut the door on the world. I hope that our daughter, Amira Solomon, now a teenager, has absorbed the example of what is involved in long-term creative projects, and goes on to make her own contributions to our world. I have been deeply fortunate to spend these years with my family.

Some of the material in this book is developed from previously published articles. In some cases there are brief overlaps in text. I have acknowledged this dependence on earlier publications at the relevant places in the book; here I acknowledge them collectively:

Solomon, M. 2014, "Evidence-based medicine and mechanistic reasoning in the case of cystic fibrosis," in *Logic, methodology and philosophy of science: proceedings of the fourteenth international congress (Nancy)*, eds. P. Schroeder-Heister, W. Hidges, G. Heinzmann, & P.E. Bour, College Publications, London, 2014, page numbers not yet available.

Solomon, M. 2012, "'A troubled area': understanding the controversy over screening mammography for women aged 40–49," in *Epistemology: contexts, values, disagreement. Proceedings of the 34th International Ludwig Wittgenstein Symposium*, eds. C. Jager & W. Loffler, Ontos Verlag, Heusenstamm, pp. 271–84.

Solomon, M. 2011a, "Group judgment and the medical consensus conference," in *Elsevier handbook on philosophy of medicine*, ed. F. Gifford, Elsevier, Amsterdam.

Solomon, M. 2011b, "Just a paradigm: evidence based medicine meets philosophy of science," *European Journal of Philosophy of Science*, vol. 1, no. 3, pp. 451–66.

Solomon, M. 2008, "Epistemological reflections on the art of medicine and narrative medicine," *Perspectives in Biology and Medicine*, vol. 51, no. 3, pp. 406–17.

Solomon, M. 2007, "The social epistemology of NIH consensus conferences," in *Establishing medical reality: methodological and metaphysical issues in philosophy of medicine*, eds. H. Kincaid & J. McKitrick, Springer, Dordrecht, pp. 167–77.

Solomon, M. 2006, "Groupthink versus the wisdom of crowds: the social epistemology of deliberation and dissent," *The Southern Journal of Philosophy*, vol. 44, Supplement, pp. 28–42.

# Contents

# Abbreviations

| | |
|---|---|
| AARP | American Association of Retired Persons |
| ACTS | Association for Clinical and Translational Science |
| AHRQ | Agency for Healthcare Research and Quality |
| CFTR | cystic fibrosis transmembrane conductance regulator |
| CONSORT | Consolidated Standards of Reporting Trials |
| CMS | Center for Medicare and Medicaid Services |
| FDA | Food and Drug Administration |
| GRADE | Grading of Recommendations, Assessment, Development, and Evaluation |
| IOM | Institute of Medicine |
| IPCC | Intergovernmental Panel on Climate Change |
| MedCAC | Medicare Evidence Development Coverage Advisory Committee |
| NGC | National Guideline Clearinghouse |
| NICE | National Institute for Health and Care Excellence |
| NIH | National Institutes of Health |
| NSF | National Science Foundation |
| OMAR | Office of Medical Applications of Research |
| STS | science and technology studies |
| USPSTF | United States Preventive Services Task Force |

# 1

# Introduction
## Beyond the Art and Science of Medicine

## 1.1 New Methods in Medicine

This book is an exploration of new epistemological approaches in medicine (called "methods" for conciseness). It focuses on consensus conferences, evidence-based medicine, translational medicine, and narrative medicine (as an example of methods from the humanities). These methods are recent—all developed within the last 40 years—although they have roots in traditional approaches such as arguments from authority, empiric medicine, causal reasoning, case-based reasoning, clinical experience, and clinical judgment. New methods typically come with methodologies, which give rationales for the appropriate use of the methods and often claim that the new methods are "transformative" or even "new paradigms" for medicine. Such claims will be looked at critically. Each of the methods can be understood historically, as developing from previous methods in response to particular problems, and each method can also be understood philosophically, in terms of its epistemic strengths, weaknesses, and interactions with other methods.

My disciplinary background is philosophy, but the book engages with a broader literature and is written for a wider audience than that of professional philosophers. Academically, this book is an example of integrated history and philosophy of science as well as an example of work in "historical epistemology" (see, for example, the works of Lorraine Daston, Uljana Feest, and Ian Hacking). It belongs in the fields of science and technology studies and (critical) medical studies. Intellectually, it is addressed to a broad academic audience, including the medical community.

Medical consensus conferences were invented at the United States' National Institutes of Health (NIH) in the 1970s. The idea was to devise a neutral process with an expert panel to resolve controversy on a topic of importance to public health. Arthur Kantrowitz's proposal for an institution of "science courts" (1967) was used as a model. Medical consensus conferences proved popular and the methods quickly spread nationally and internationally. In 1990, an Institute of Medicine (IOM) report on the international use of consensus conferences began by stating: "Group judgment methods are perhaps the most widely used means of assessment of medical technologies in many countries" (Institute of Medicine 1990b).

Just two years later, however, Gordon Guyatt and his colleagues at McMaster University in Canada declared that "A new paradigm for medical practice is emerging" (Evidence-Based Medicine Working Group 1992), and with this they announced the birth of evidence-based medicine. Evidence-based medicine has been prominent since the mid-1990s, producing assessments of medical interventions such as pharmaceuticals, surgery, and diagnostic tests. Specific kinds of assessments in evidence-based medicine include judgments of trial quality, systematic evidence review, and meta-analysis. Group judgment—so central for consensus conferences—is typically not counted as evidence or it is ranked as the "lowest quality" in a hierarchy of evidence. Consequently, group judgment is used with reluctance in coming to conclusions in evidence-based medicine. Evidence-based medicine spread from its beginnings in epidemiological medicine in Canada and the UK to become a worldwide movement with important institutions such as the Cochrane Collaboration and the Agency for Healthcare Research and Quality (AHRQ)'s evidence-based practice centers (Daly 2005). David Eddy writes, "By the end of the 1990s it was widely accepted that guidelines should be based on evidence, and the only acceptable use of consensus-based methods was when there was insufficient evidence to support an evidence-based approach" (2005: 15).

In 2005 Donald Berwick, one of the early champions of evidence-based medicine,[1] lamented that we have "overshot the mark" and

[1] Donald Berwick is better known for his participation in the medical decision-making movement, but he is also well known for his endorsement of evidence-based methods.

created "intellectual hegemony" with evidence-based medicine (Berwick 2005). Berwick recommends "broadening" evidence-based medicine with a more "pragmatic science" in which we use local knowledge and less formal methods in the initial stages of research development. In his view, clinical medicine has become dominated by epidemiological methods (which are the tools of evidence-based medicine), and in the process has lost sight of the kinds of scientific reasoning involved in the discovery and development of theories and interventions. His fear is that evidence-based medicine will lead only to conservative innovation and be limited to interventions ready to go into clinical trials. Bolder and more creative interventions may require tinkering, small initial studies, and careful clinical observation, rather than large-scale randomized controlled trials, at least at the outset. What Berwick is recommending is part of "translational medicine,"[2] first regularly mentioned in the literature around 2003 and now the focus of new national and international initiatives (Woolf 2008). In 2006 the NIH started its new Clinical and Translational Science Awards and has, so far, funded 62 research institutes for translational medicine at a cost of more than half a billion dollars annually. In the UK, the National Institute for Health Research has funded 11 centers for translational medicine (recently named Biomedical Research Centers) at a cost of 800 million pounds over five years and the European Advanced Translational Research InfraStructure in Medicine recently received 4.2 billion euros (over three years) from the European Commission.[3]

Alongside these developments, there has been a reaction to evidence-based medicine and more generally to "scientific medicine" in some parts of the medical profession and in the medical humanities. Both medicine guided by consensus conferences and evidence-based medicine are commonly derided as "cookbook medicine." From the physician's perspective, the general clinical guidelines that issue from consensus conferences and evidence-based medicine are said to devalue more nuanced clinical judgment and expertise and to interfere with physician autonomy. From the patient's as well as the physician's perspective, the criticism is that clinical guidelines—based on either evidence-based

---

[2] Specifically, it is the most prominent part of translational medicine, known as T1 research, or "bench to bedside and back."

[3] These figures were last checked in May 2014. They may have changed since then.

medicine or consensus conferences—do not treat patients as individuals and devalue the importance of the physician–patient relationship. There is also concern, especially in the US, that such clinical guidelines are the first step toward increased medical regulation by bureaucratic organizations such as the government, health insurance companies, and the legal profession. Despite protestations from the evidence-based medicine community, and even restatements of the goals of evidence-based medicine to include such friendly phrases as "the care of the individual patient" (Sackett et al. 1996), these criticisms persist. In the academic community, they have fostered the growth of the medical humanities, especially the field of narrative medicine (Brody 2003 [1987], Charon 2006a, Montgomery 1991, Zaner 2004) and the overlapping field of phenomenology of medicine (Carel 2008, Leder 1990, Pellegrino et al. 2008, Toombs 2001).

## 1.2   Shortcomings of an Art versus Science Analysis

On the surface, it appears that what is going on is scientific change on the "science" side of medicine, emphasizing different methodologies over time (first consensus conferences, then evidence-based medicine, then translational medicine), and a reaction or response from the "art" side of medicine, mostly concerned with insisting that medicine requires not only scientific knowledge but also implicit and/or humanistic skills. Kuhn's (1962) concept of a "paradigm" is often mentioned,[4] not only for methods on the "science" side but also for the methods on the "art" side—such as narrative medicine—where Rita Charon, for example, has claimed that she is putting forth "A new philosophy of medical knowledge" (2006a: 39). In taking a closer look, however, I find that analysis in terms of scientific paradigm change on the "science" side and reaction to the scientific image on the "art" side gives an incomplete and misleading picture of the epistemic situation. Here are some of my concerns.

---

[4] Kuhn (1962) introduced the term "paradigm" with a specific set of meanings about revolutionary change in science. The term quickly entered common usage, where it marks any kind of major change in theory or practice, and in particular was used to mark the introduction of evidence-based medicine. As Kenneth Bond pointed out to me, there is typically little understanding of the details of Kuhn's ideas in these uses of the term.

While the methods of consensus conferences, evidence-based medicine, narrative medicine, and translational medicine exhibit some of the characteristics of Kuhnian paradigms (such as dependence on a set of core exemplars, claims of universal applicability, and claims to be transformative), they do not evince a Kuhnian narrative of paradigm change. There has been a change in focus from consensus conferences to evidence-based medicine to translational medicine, but new methods do not replace older methods in the manner that Kuhn (1962) describes for historical examples such as change in physical theory during the Scientific Revolution. (In that example, Aristotelian physics was completely overthrown in favor of Newtonian mechanics, and a new mathematical-experimental method replaced earlier observational methods.) Group judgment is still essential in medicine: consensus conferences have diminished in frequency since the 1980s but still continue (although the NIH Program has recently ceased). Ironically, group consensus methods have recently been proposed by the GRADE (Grading of Recommendations, Assessment, Development, and Evaluation) Working Group to resolve uncertainty about the proper hierarchy of evidence in evidence-based medicine (Guyatt et al. 2008), suggesting that expert consensus partly constitutes the foundation of evidence-based medicine. Evidence-based medicine continues to be strong, despite the recent popularity of translational medicine and despite criticisms from the medical humanities.

Describing the situation as one in which the "science" and the "art" of medicine are in tension also gives an incomplete characterization. Some of the methods, most notably the group judgment of consensus conferences, do not fall on either the "art" or the "science" side. The authority of group judgment comes in part from the ideal of rational deliberation of experts, and rational deliberation is used in a range of political as well as scientific settings. Group consensus is often thought of as an epistemic ideal, although not as an ideal specifically for science. Moreover, scientific approaches such as causal (pathophysiological and/or mechanistic and/or case-based) reasoning, evidence-based medicine, medical decision making, and translational medicine all differ from one another. There is more than one way of being scientific, and each way is of epistemic interest. Focusing on the "art" and "science" dichotomy can obscure this.

A distinction between the "art" and the "science" of medicine is long-standing, although it has taken on a new resonance since the birth of what we call "scientific medicine"[5] in the late nineteenth century. It probably has roots in the Greek distinction between *techne* and *episteme*, that is, applied knowledge or craft versus theoretical knowledge.[6] Medicine requires both *techne* and *episteme* because it requires both clinical skills and understanding of a systematic theoretical framework: biology. The dualism (both "art" and "science") assumes that something like the Greek distinction between different kinds of knowledge is still useful. I will question that usefulness.[7]

The historian of medicine Charles Rosenberg focuses on a modern version of the art/science dualism, one between "holism" (on the art side) and "reductionism" (on the science side), to explain the epistemic tensions in twentieth-century medicine (2007). "Reductionism" is a kind of medical explanation that became particularly popular after the birth of "scientific medicine" in the late nineteenth century, when disease was increasingly understood in terms of organs, parts of organs, microorganisms, and biochemical processes rather than in terms of the whole organism (as in humoral theories) or its environment (such as miasmas). There was a conservative reaction within the medical community to scientific medicine (also sometimes called "the biomedical model") and this was expressed in the contrast between cure (from the "science" of medicine) and care (from the "art" of medicine).

In the early to mid-twentieth century, scientific medicine increased its resources with the development of antibiotics, the design of the randomized controlled trial, and advances in transplantation, critical care, surgery, and oncology. The rise of genetics added to the options for reductionist explanations. A little later, the "art" of medicine also developed further with the establishment of medical ethics as an academic field, the idea of the "biopsychosocial" model for clinicians (Engel 1977), a new

---

[5] I keep this term in quotation marks in order to avoid begging the question about the scientific character of medicine.

[6] Greek epistemology is of course complex and varied; other epistemic concepts such as *phronesis, theoria, poiesis, praxis, logos,* and *ethos,* play important roles. It may be that the art/science bifurcation never sufficed as a classification of methods.

[7] I am not the only one to challenge the usefulness of the art–science distinction for medicine. Kenneth Goodman (2003) does so also, mostly because he thinks that the distinction presupposes an outdated philosophy of science. I agree with his analysis and offer the same and additional reasons for challenging the usefulness of the art–science distinction for medicine. Goodman's work has not yet received the attention it deserves from epistemologists of medicine.

emphasis on equality in the doctor–patient relationship (Katz 1984), a focus on the need to address the patient's experience of suffering (Cassell 1976, Cassell 1991), the popularity of "holistic" health approaches, and the growth of the medical humanities.

In my assessment,[8] the recent distinction between the "art" and "science" of medicine presumes a traditional logical positivist or logical empiricist philosophy of science. Such philosophies of science dominated between the late nineteenth century and the 1960s. They hold that scientific knowledge is comprised of general laws (and/or general mechanisms) which are applied to specific circumstances by adding precisely stated factual assumptions and using logical rules of inference. They also hold that science is driven by cognitive goals such as truth and predictive power and insist that political values and individual preferences should play no role in theory choice. This traditional view about science was superseded in the post-Kuhnian period. Kuhn explored the messiness of science and the imprecise nature of scientific judgment in his "Objectivity, Value Judgment and Theory Choice" (Kuhn 1977). It is now widely recognized that science—even a so-called "mature" science like physics—is not free of intuitive, imprecise, narrative, and evaluative elements.[9] Science is messy (Collins & Pinch 1993). One might say that actual science is not really a "science" as science has been traditionally understood to be: i.e. precise, value free, and fully explicit. This has recently been acknowledged specifically for evidence-based medicine by Michael Rawlins, who chairs the National Institute for Health and Care Excellence (NICE) in the UK. In his Harveian Oration (2008), Rawlins frankly acknowledges the importance of "judgment," by which he means those parts of methodology that cannot be stated as formal and precise rules. Such "judgment" is part of science, and no area of science can do without it.

Sometimes the bifurcation of methods is expressed as "hard" versus "soft" medicine, sometimes as the "precise" versus the "messy," sometimes as "reductionistic" (usually meaning organ-based, cell-based, biochemical, or genetic) versus "holistic" approaches (as in Rosenberg's analysis), sometimes as Western versus Eastern approaches, and so on.

---

[8] Kenneth Goodman (2003) makes a similar assessment.
[9] A few still argue for a traditional or slightly modified logical empiricist view. This book is not about that controversy.

Of course, the details of these opposed concepts are different from one another, but what is interesting is that the discourse for each opposing pair of methods is so similar. People often have a personal hierarchy, preferring either the "science" or the "art" side. Not surprisingly, there is gender ideology expressed in the bifurcation of methods, with the "science" side (including terms such as "hard," "precise," "reductionist," "Western," "objective," "rational," "general," "abstract") gendered masculine and the "art" side (including terms such as "soft," "vague/messy," "holistic," "Eastern," "subjective," "intuitive," "particular," "concrete") gendered feminine.[10] Evidently, this conceptualization of the epistemology of medicine is expressing something of importance to its users. It also harks back to C.P. Snow's famous description of "two cultures" (1959) now applicable to two kinds of knowledge within medicine. For some, the appeal of medicine comes in part from an understanding of the field as a "marriage of opposites" because they take pride in their own abilities to make use of seemingly divergent methods.

Those who are familiar with the tools and language of post-structuralism will notice that, in positioning myself to reject this traditional framework, I am proposing to deconstruct a binary set of concepts ("art" versus "science"). I am happy to be able to do this particular deconstruction and acknowledge that it may serve wider goals. My own goal is to offer a new and somewhat general way of conceptualizing and evaluating epistemological techniques in medicine.[11]

Since the development of evidence-based medicine in the 1990s, the dichotomy between the art and the science of medicine has, if anything, deepened. Evidence-based medicine has continued to be criticized from the "art" side as "cookbook medicine,"[12] which denigrates clinical expertise and does not straightforwardly apply to individual patients. The biopsychosocial model—an early attempt to connect the "art" and the "science"

---

[10]   John Clarke remarked to me that in surgical contexts "concrete" is gendered masculine and "abstract" is gendered feminine. I am more familiar with philosophical contexts. Concepts of gender have some fluidity.

[11]   I do not emphasize this way of framing my work because I do not wish to ally myself with all of its details, such as the relativistic epistemology, social constructivist ontology, and avoidance of all generality and normativity typically adopted along with opposition to binary concepts (e.g. by Derrida, Foucault, Kristeva, and Butler).

[12]   The term "cookbook medicine" insults most cookbooks as well as consensus conferences and evidence-based medicine. (Cookbooks mostly offer suggestions and ideas, not "one size fits all" or "algorithms.")

of medicine—has faded,[13] to be replaced by narrative medicine. Narrative medicine is a new approach in medical ethics and medical humanities (Brody 2003 [1987], Charon 2006a, Zaner 2004), in which the techniques of literary analysis are used to improve the practice of medicine. Phenomenological approaches (from Continental philosophy) are also popular (Carel 2008, Leder 1990, Toombs 2001, Toombs 1992) and claim to provide a discussion of the experience of illness that "the scientific perspective" cannot provide and that doctors should know. "Holism" has expanded to become "integrative medicine." From the "science" side, clinical experience has sometimes been denigrated by evidence-based medicine, and medical humanities has often been treated as peripheral to medical training and practice. There have been some attempts to bring the "art" and the "science" together (e.g. Meza & Passerman 2011), but these still rely on maintaining a dichotomy of methods.

The changes in medical methods that have taken place in the last 40 years also raise questions about progress. How different are the new methods from older methods? For example, is evidence-based medicine different from Claude Bernard's "experimental medicine"? Are consensus conferences different from more informal expert agreement? Is narrative medicine different from the traditional adage "listen to the patient"? And does translational medicine include any new methods at all? How do we see beyond the catchphrases, buzzwords, and rhetorical stances of these new medical methods to figure out what their real contributions are? My answers to these questions will be developed through detailed assessments of each methodology. It will turn out that some of the changes are epistemically progressive, while others are less so.

## 1.3  Current Literature on Methods in Medicine

Recent book-length works in the epistemology of medicine focus on just one method at a time. Harry Marks's *The progress of experiment: science and therapeutic reform in the United States, 1900–1990* (1997)

---

[13] Dan Sulmasy (2002) has continued with the biopsychosocial model, developing it into the biopsychosocial-spiritual model. But the model has not received much attention, and narrative medicine is more likely to be used for teaching the "art" of medicine to medical students. I will discuss this in Chapter 8.

looks at the development of therapeutic research in the twentieth century, and focuses on the evolution of the randomized controlled trial, which eventually became the core of evidence-based
medicine. Stefan Timmermans and Marc Berg's *The gold standard: the challenge of evidence-based medicine and standardization
in health care* (2003), Jeanne Daly's *Evidence-based medicine and the
search for a science of clinical care* (2005), and Jeremy Howick's *The
philosophy of evidence-based medicine* (2011b) describe the development of evidence-based medicine. Kathryn Montgomery's *How
doctors think: clinical judgment and the practice of medicine* (2006)
explores the intuitive judgments of experienced physicians. Rita
Charon's *Narrative medicine: honoring the stories of illness* (2006b)
looks at how the techniques of literary analysis can improve both
diagnosis and patient care. Jodi Halpern's *From detached concern to
empathy: humanizing medical practice* (2001) explores the role of professional empathy in patient care, and Havi Carel's *Illness: the cry of
the flesh* (2008) develops a phenomenological approach. (There are no
comprehensive or book-length studies of medical consensus conferences or translational medicine.)

These works tend to be either carefully neutral social science accounts
of an epistemic tradition (Marks 1997, Timmermans & Berg 2003) or
enthusiastic promoters of new methodologies (Daly 2005, Charon 2006a,
Carel 2008, Howick 2011b, Montgomery 2006, Halpern 2001). There are
normative philosophical accounts (i.e. accounts which examine the
epistemic advantages and disadvantages of a particular methodology)
only of the epistemic limitations of randomized controlled trials and
evidence-based medicine (e.g. Howick 2011b, Bluhm 2005, Cartwright
2007a, Sehon & Stanley 2003, Howick 2008, Tonelli 1998, Worrall 2002,
Worrall 2007b). There is no substantial study of the ways in which the
different methodologies fit together, react to one another, sometimes
disagree with one another, and are negotiated in the context of specific
research and clinical questions. This book is intended to fill these gaps
in the literature, and to address the normative questions more fully.
Philosophers are sometimes too quick to come to normative conclusions;
I strive to take the time to understand the full epistemic context before
making any assessments.

# 1.4   My Approach: Naturalistic, Normative, Applied, Pluralist, Social Epistemology

Since the 1960s and the work of Thomas Kuhn (1962), Michael Polanyi (1958), Norwood Russell Hanson (1958), Mary Hesse (1966), Stephen Toulmin (1958), and others, the imprecise, messy, and non-logical (such as analogical and narrative) characteristics of all scientific methods have been much better appreciated, at least by philosophers of science and others working in science studies. I bring this improved understanding of science to the analysis of methods and methodologies in medicine. When science is more fully understood, I argue, there is no usefulness in positing a supplementary "art" of medicine to balance and complete the methodology. Another way to state this is to say that we have learned that science is not a "science" in the sense understood by traditional (positivist) philosophers of science, so we have no need for a separate category of "art" in the sense understood by traditional philosophers of science either. The science/art dichotomy is no longer a fruitful disciplinary divide. It is an intended casualty of this study. I look at each method on its merits, bypassing the traditional dichotomous classification into science or art.

My philosophical approach is that of naturalistic, normative social epistemology. This means that is also *applied* epistemology. I do not attempt to rebuild our practices from the ground up or build abstract (typically partial or idealized) models of them, but rather I strive to make concrete suggestions for improvement based on a careful historical and contextual analysis of where the practices are now. The tools for doing this convincingly are interdisciplinary, and both the approaches and the results should be relevant, congenial, and comprehensible to a wide range of scholars and researchers in the medical sciences and humanities.

It is especially challenging to attempt normative assessments in medicine because different methods sometimes have different goals. For example, humanistic approaches sometimes claim to be "caring for the whole person" rather than "treating the disease." The latter, they claim, is the target of more scientistic approaches. Eric Cassell (1976, 1991) has drawn attention to this distinction, elaborating it by claiming that the physician's duty is to try to "heal" rather than merely cure, and to aim at the goal of "the relief of suffering" rather than the goal of normal physical functioning. Many others have similar views (e.g. Charon 2006a, Carel 2008, Engel 1977, Montgomery 2006, Halpern 2001, Greer 1987, Riesman 1931, Sulmasy 2002). Of course it follows that evaluations of the various different methods will depend on which goals are adopted. There is also room for debate over how much to value some goals, and when to expect health care professionals (rather than some other helping professions) to satisfy them. I will attempt to make normative assessments with this complexity in mind.

Analyses of methods and methodology in medicine have practical consequences. For example, clinicians reading this book can gain more awareness of what it means to use "clinical judgment" to depart from the recommendations of evidence-based medicine, and when such departure is and is not justified. They can explore the ways in which favored methods may be fallible, or may be ritualistic or arbitrary, which makes it possible to use the methods less dogmatically and more judiciously. For methodologists—investigators who want more understanding of the methods and how they fit together and change over time—it will be possible to step back from the rhetoric that surrounds each method to learn its actual history and its achievements; to explain, and perhaps even predict, newer trends such as translational medicine. This can facilitate more reflective, rather than reactive, research policies.

I draw novel conclusions in my analysis of each method. Here, briefly, are some of them so that the reader can see where I am headed.

NIH Consensus Conferences (and other "technical" consensus conferences) can best be understood as *epistemic rituals* in which the process matters more than the product and the *appearance* of objectivity, democracy, fairness, and expertise matters more than the reality of freedom from bias. They perform an epistemic function similar to traditional authority-based knowledge, although they do so in a new, more tightly choreographed, and in some ways more democratic manner. These consensus conferences are better designed to disseminate knowledge than they are to produce it, although they have difficulties with dissemination also. Other consensus conferences, especially those based on the "Danish" and "interface" models, are more *political rituals* that democratically negotiate joint action.

Evidence-based medicine has roots in Greek "empiric medicine," but is also the product of twentieth-century methodological advances in statistical techniques. It provides powerful techniques for assessing medical technologies that have modest to moderate effects. (It is too fussy—methodological overkill—for assessing medical technologies with consistent and large effect size.) It has core successes that serve as exemplars in the manner of a traditional Kuhnian paradigm: A. Bradford Hill's 1948 trial of streptomycin for tuberculosis functions much like the case of the inclined plane for Galileo's mechanics. Evidence-based medicine is not a universal or a complete research methodology, and it is much less accurate in practice than generally expected (Ioannidis 2005, LeLorier et al. 1997).

Translational medicine is, in large part, a predictable corrective to the methodological limitations of evidence-based medicine and not as novel as the neologism suggests; it has roots in case-based reasoning, clinical judgment, and general causal reasoning.

And finally, humanistic methods such as narrative medicine are at best thoughtful and expansive responses, and at worst reactionary responses, to the increased importance of both evidence-based medicine and reductionist (genetic, neural, and biochemical) explanations. They express a wish to hold on to the intimacy of the traditional physician–patient dyad as the core of the practice of medicine. This wish is familiar and widely shared, but it is instructive to look at alternatives.

Each of the methods I discuss has something *obvious* about it, and each has something *odd* about it. For example, rational deliberation to consensus is a common understanding of objectivity (especially in political contexts), yet it seems strange to expect 10–20 physician-researchers seated around a table to come to a worthwhile consensus in 48 hours—with no additional research—on a scientifically controversial matter. Trisha Greenhalgh (as quoted in Howick 2011b: 18) irreverently describes the method of expert consensus as "GOBSAT": "Good Old Boys Sat Around the Table." Jonathan Moreno (1995) has also commented critically on the weight given to expert consensus. Evidence-based medicine at first seems to state the obvious—that medical knowledge is based on evidence—yet it has a precise (some think narrow, and some think refined) view of what counts as quality of evidence. "Translational medicine" seems to be a new term for old practices of causal reasoning and trial and error intervention. Narrative medicine emphasizes listening to the patient, which is widely regarded as good basic medical practice, yet also claims that listening with the tools of literary analysis, rather than commonsense attention, gives the physician the most information.

I selected these four methods for attention because of this dual and paradoxical epistemic character (obviousness and oddness), expecting—and finding—that this means that something of epistemic interest is occurring. I pay particular attention to the areas in which the methods can lead to different conclusions on the same medical topic. These are sites of potential controversy, in which we can see the relationships between different methods. We see when and how one method can be favored over another. This opens up a new level of potential criticism, not of the methods themselves, but of the possible hierarchical relationships between different methods. I will argue *against* the usefulness of a meta-methodology that ranks the methods; controversies are best worked through by taking each method further rather than by trying to disregard some of them. It is also notable that although new methods come into vogue, older methods do not disappear. Behind the typical rhetoric of excitement about the newest method(s) is a genuine pluralism of available methods in which the choices increase as new methods are developed.

If each method had its own domain of application, and the domains did not overlap, we would have what I call a "tidy pluralism" (such as is found for theories in Longino 2002 and Giere 2006). We would choose the appropriate method for each problem, and then assemble all the results into a coherent whole. But methods in medicine do not form a tidy pluralism (nor do methods in other areas of science and technology). Sometimes more than one method is appropriate for a problem, and sometimes those methods yield different results, a situation with some incoherence, that I call "untidy pluralism." For example, causal reasoning and clinical expertise/judgment often conflict with evidence-based medicine. One of many examples is the continuing practice of using a fetal heart monitor during labor because of "clinical experience/judgment," despite no evidence from clinical trials that it improves outcomes. Another example is the use of vertebroplasty for painful osteoporotic vertebral fractures, which makes sense physically, but is not supported by randomized controlled trials (Buchbinder et al. 2009). Consensus conferences can make recommendations that conflict with systematic evidence reviews, as has occurred several times for recommendations for screening mammography (see Chapter 9). Narrative approaches can conflict with causal models, as happens when a patient's story—for example, about the reason they got breast cancer—disagrees with plausible physical accounts (see Chapter 8).

Equally interesting are those cases where conflict might be expected, but none occurs, either because of prior structural decisions about which methodology should be privileged, or because of some other strategy to manage disagreement. For example, after 2000, NIH Consensus Conferences were required to take place only *after* the panel received a systematic evidence review from the AHRQ; this evidence review anchored the panel deliberations. Since the panel was charged with making only evidence-based recommendations, they could not end up with much disagreement with the evidence review. In such cases, evidence-based medicine was structurally privileged and conflict was avoided. Another example is that clinical guidelines based on evidence-based medicine often include a qualification that the practitioner should use his or her "clinical judgment" in deciding whether and how to apply the guidelines. This qualification gives some privilege to clinical judgment and allows practitioners to override recommendations from evidence-based medicine, usually over a range of "atypical" cases.

I argue *against* devising or implementing a general hierarchy of methods—a meta-methodology—although that would be a simple and unambiguous way to resolve all conflict. Since all the methods are fallible, both in theory and in practice, and none are highly reliable, none should be given a "trump card" over others. There is no epistemic justification for a general hierarchy.

My approach will be two-pronged. First, the careful analysis I give of each method will reveal areas in which the method is particularly strong, and others in which it is less strong or even weak. For example, clinical judgment is particularly susceptible to availability and salience biases. Consensus processes are susceptible to groupthink. Evidence-based medicine is susceptible to publication bias, time to publication bias, and biases created by the interests of pharmaceutical company sponsors. Translational medicine has a high expected failure rate in the so-called "valley of death."[14] Narrative methods are subject to the "narrative fallacy," in which narrative coherence can obscure as much as it can reveal. My assessments take the particular strengths and weaknesses of each method into account. Secondly, conclusions should be drawn only provisionally because we cannot completely figure out methodological

---

[14] This is the pharmaceutical companies' term for failure to show safety, efficacy, or effectiveness at some stage of drug development. See Chapter 7.

questions in advance of engaging the world. Surprises are to be expected, when some methods may show unusual success in an unlikely domain and others may fail despite high expectations. Methods are tools for engaging with the world, but, as Donna Haraway writes (1991: 199), "we are not in charge of the world." Our methods develop along with our theories and our successful interventions.

Epistemological humility and what I call "epistemological equipoise" are appropriate attitudes for the methodological pluralist in medical research and clinical practice. (Epistemological equipoise is inspired by the well-known term "clinical equipoise," which is used widely in medical ethics contexts to mark situations in which treatments under consideration are, prior to testing, estimated as equivalently effective by experts.) Epistemological equipoise requires aiming for as much methodological impartiality as possible. This is not achieved by adopting a neutral stance (as if that were possible) but rather by inhabiting each method as well as possible, listening to the perspectives of its advocates. Epistemological humility is also required because methodological equipoise is so difficult to attain.

## 1.5  Disrespecting the Disciplines

If it is no longer helpful to describe medicine as "both an art and a science," how should we talk about its methods? This book aims to get beyond the "art versus science" taxonomy in order to gain a richer understanding of recent methods. My goal is not to classify them anew in some more updated disciplinary framework, but to improve our uses of the methods. My goals are pragmatic rather than taxonomical.

More generally, I believe that disciplinary boundaries deserve some disrespect at this time, the beginning of the twenty-first century, when interdisciplinary research still struggles for recognition and resources despite its recent track record of innovation. The social and institutional structures of the traditional disciplinary boundaries impede interdisciplinary work, especially for younger scholars still looking for an academic home.

Medicine does not need to be classified as an art, a science, both, or anything else to do work in the epistemology of medicine.[15]

---

[15] "Technoscience" (see note 18, this chapter) is probably the most accurate of our current categories, because it captures the ways in which medicine makes use of biological

## 1.6  Selection of Methods for Discussion

This book draws on work from several literatures and disciplines. It is a "big picture" account that is grounded in detailed historical and philosophical exploration of particular methodological traditions. This kind of work needs a balance between comprehensiveness and manageability. I plan to use considerable detail in sufficient examples to achieve my aims, but I will not be able to look at all methods in medicine or at all the historical and philosophical details of the cases that are epistemically interesting. Also for these reasons of manageability, I will give only brief accounts of traditional methods in medicine, such as case study reasoning, appeal to authority, clinical experience, and causal-mechanistic knowledge, and I will say only a little about other new methods (or approaches) such as medical decision making, shared decision making, systems analysis, realist review, personalized medicine, integrative medicine, and hermeneutical medicine. The same approaches can be used in investigating these methods (or approaches) also, and I hope to do this in future work.

I have selected four methods for detailed discussion—consensus conferences, evidence-based medicine, translational medicine, and narrative medicine—based on my assessment of their substance and impact. Other methods will be discussed also, but briefly.

## 1.7  Summary of Major Themes in Recent Philosophy of Science

Philosophy of medicine, as other branches of philosophy, is sometimes thought to be overly abstract and irrelevant to practical questions. This criticism developed in the mid-twentieth century, when the tools of logical and conceptual analysis were the primary methodology for philosophy of science. I am bringing a more recent, more empirically engaged literature in philosophy of science to questions in the epistemology of medicine. This literature is more appreciative of the implicit, imprecise, contextual, intuitive, and unpredictable characteristics of scientific practice, as well as the qualified and fallible nature of scientific achievements.

knowledge to satisfy human goals. Nevertheless, my strategy in this book is to avoid disciplinary classifications altogether. Perhaps at some time in the future they will become interesting again.

The following are eight specific features of this post-Kuhnian literature that influence my work:

1. There is more than one scientific method (methodological pluralism). See, for example, the work of William Bechtel (1993), Lindley Darden (2006), Helen Longino (2002, 1990), and Alison Wylie (2000).

2. Science is a social product, rather than the product of individual scientists working in isolation. Examples are the work of David Hull (1988), Helen Longino (2002, 1990), Miriam Solomon (2001), and Paul Thagard (1993).

3. History of science (as well as sociology of science, rhetoric of science, and other science studies disciplines) is essential for doing contextually sensitive philosophy of science. Philosophy of science can also guide research questions in other science studies disciplines. (As such, I advocate "integrated history and philosophy of science," and indeed, integrated science studies.) Examples are the work of Hasok Chang (2004) and Lindley Darden (2006).

4. Science is influenced by values, politics, and interests. See the work of Heather Douglas (2009), Ronald Giere (2006, 1988, 1999), Sandra Harding (1991), David Hull (1988), Philip Kitcher (1993, 2001) Helen Longino (2002), and Alison Wylie (2002).

5. Scientific objectivity needs rethinking, since it can no longer be conceptualized in the traditional Baconian manner as pure, unbiased thought. Some new conceptualizations are given by Sandra Harding (1993), Helen Longino (2002, 1990), and Miriam Solomon (2001). In analytic epistemology, the reliabilism[16] of Alvin Goldman (1992, 1999, 2002) has been an influential alternative account of objectivity that does not require reasoning to be "bias free."

6. Science is a practice in which some of the constituents are tacit. Not all of science is expressed linguistically or mathematically, see Ian Hacking (1983), Thomas Kuhn (1962, 1977), and Michael Polanyi (1958).

7. Scientific theories are perspectives on the world, rather than complete representations, and more than one perspective may be useful

---

[16] "Reliabilism" is a technical term created by Alvin Goldman to describe his theory of justification. For Goldman, a justified method is a reliable method (one that gets to the truth more often than not).

in particular domains. This is referred to as ontological pluralism.[17] See Nancy Cartwright (1983, 1999), John Dupré (1993), and Ronald Giere (2006, 1999).

8. Science and technology can often be considered together, as "technoscience."[18] This follows from recognition of the role of values and interests in the development of science, and from the importance of technological success (success in manipulating the world) in scientific success.

This literature in philosophy of science is of course influenced by important work in the more general interdisciplinary field of science and technology studies (STS), such as the work of Karen Barad (2007), Harry Collins (1985), Collins and Evans (2007), Peter Galison (1987, 1997), Donna Haraway (1991, 1989), Sheila Jasanoff (2004, 2005), Evelyn Fox Keller (1985), Bruno Latour (1993, 2005), Andy Pickering (1984, 1995), and Steven Shapin (1994, 1996, 2008).

The tools of recent philosophy of science and STS have not often been applied to recent medical epistemology. Notable exceptions include work by Nancy Cartwright (2007a), Rebecca Kukla (2005), and Stefan Timmermans with Marc Berg (2003). This book will add to these works, bringing recent theoretical tools to address new questions in philosophy of medicine.

## 1.8  Research Methods and Materials

The historical record for each new method in medicine is different, and I have adapted my own research methods accordingly. There is certainly room for much more historical research, including archival research, which would offer finer-grained accounts than I am able to give in this broader and philosophically motivated inquiry, and which would perhaps also correct some of my claims. Here is a brief overview of the methods and materials I use to investigate specific methods in medicine.

---

[17] Sometimes the position is simply referred to as "pluralism," but I prefer using the full term "ontological pluralism" in order to distinguish the position from methodological pluralism. That is, pluralism about entities (or theories) is different from (although often complementary to) pluralism about methods.

[18] The term "technoscience" was coined by Gaston Bachelard (1953) and is now widely used in science studies.

For consensus conferences, my resources consist of online materials from the NIH Consensus Development Conference Program (historical and current); the literature on NIH and other consensus conferences (typically in the form of journal articles in medical publications); informal interviews with past directors and deputy directors of the Office of Medical Applications of Research (OMAR) (Dr. John Ferguson, Dr. Barnett Kramer, Dr. Susan Rossi); formal reports evaluating the NIH Consensus Development Conference Program such as a 1989 Rand Corporation report, a 1990 IOM report, the 1999 Leshner report, and other reports; formal evaluation of consensus conferences outside of the US in another 1990 IOM report; and my own experience attending the public portions of the 2004 NIH Consensus Conference on celiac disease. Because no one else has worked on the history of medical consensus conferences, my account is particularly detailed.

These historical materials on medical consensus conferences are brought together with a critical philosophical analysis of claims that rational consensus is brought about through the deliberation of a group of experts. The analysis will include an engagement with the general philosophical literature on deliberation and consensus.

More has been written about evidence-based medicine than has been written about all the other methodologies put together. Jeanne Daly (2005) has written a careful history of the movement; Timmermans and Berg (2003) have written a sociological account of the importance of evidence-based medicine for professional power and standardization of care; and Jeremy Howick (2011b) has published a philosophical account of evidence-based medicine. Kenneth Goodman has written an insightful text on the implications of evidence-based medicine for medical ethics (2002). Numerous textbooks give instruction in the methods of evidence-based medicine. Online resources such as the Cochrane Collaboration and the AHRQ evidence-based practice centers provide specific examples of evidence-based medicine. There is a substantial literature expressing reservations with the methods and results of evidence-based medicine, mostly in medical journals and some in philosophical journals. A common criticism is the claim that evidence-based medicine is "cookbook medicine" which does not treat patients as individuals and devalues clinical judgment. Other criticisms focus on the assumed lack of bias in the randomized controlled trial, challenges to hierarchies of evidence, the absence of evidence for evidence-based medicine, the scientific inadequacy of evidence-based medicine, and the unreliability of evidence-based medicine. The challenge here is to make sense of a wide range of criticisms of evidence-based medicine, and to use the best of these criticisms, along with my own reflections, to make normative recommendations. This work in applied epistemology uses both logical and empirical tools.

Translational medicine is an emerging approach to medical research. It is unclear how it will develop and how much success it will have, but there is already much to discuss about its origins and its important methodological characteristics. The approach I will take is to look at both the literature on translational medicine (which is in the form of journal articles in medical publications) and at the information disseminated by funding organizations supporting translational research, for example the NIH Roadmap for Medical Research (2004), the NIH Clinical and Translation Science Awards announcements (since 2006), the European Commission, and the National Institute for Health Research in the UK. Two important journal articles on translational medicine are Woolf (2008), which explores two kinds of translational medicine (called T1 and T2), and Berwick (2005), which suggests (in terminology that antedates "translational medicine") some reasons why translational medicine follows an era of evidence-based medicine.

In the medical humanities, a variety of related methods, such as professional empathy (Halpern 2001), narrative expertise (Charon 2006a), phenomenological description (Carel 2008, Toombs 2001), and even clinical judgment (Montgomery 2006) are advanced as necessary supplements to "scientific" methods. Narrative medicine is perhaps the best known; a new master's program directed by Rita Charon began at Columbia University in fall 2009, and the International Network for Narrative Medicine was created by Rita Charon and Brian Hurwitz in June 2013. I cast a wide net in looking at the literature on narrative medicine, and include the literature on related humanistic approaches. Materials are extensive, and in the form of both books and articles. This medical humanities literature is distinct from the medical ethics literature (even though some medical ethics is narrative ethics). The claim of the medical humanities literature is that medical care can have improved outcomes with humanities input rather than (or in addition to) claims that medical care will be ethically better with humanities input. I view the humanities methods as related, despite their disparate origins, because of their common reliance on the patient's experience of illness, the need for physician empathy (or some kind of emotional engagement), the emphasis on individual cases and stories, and the reliance on the intuitive, inexplicit skills of the physician. As mentioned in Section 1.4, one of the challenges in assessing these methods is that they define their health care outcomes (e.g. "healing") in ways that are incommensurate with some specific endpoints of evidence-based medicine (e.g. "viral load below 50"). I handle this by acknowledging the pluralism of goals, while at the same time being open to the possibility that different methods can often contribute toward the same goal. For example, evidence-based medicine may be able to address suffering (the development of appropriate metrics is beginning[19]) and narrative methods can aid in accurate diagnosis.

## 1.9   Outline of Chapter Contents

Here is a brief outline of the chapters in this book. Chapters 2–4 are on medical consensus conferences. Chapter 2 explores the history and epistemology of the NIH Consensus Development Conference Program, which was the original model for all medical consensus conferences.[20] The epistemic ambiguities, ironies, and compromises in the NIH Program set the stage for the ways in which the model was adapted both nationally and internationally to fit different contexts. Chapter 3 looks at a variety of medical consensus conference programs, and shows how they developed differently in the US and Europe in ways that reflect national ethos as much as epistemological concerns. Chapter 4 is a general philosophical exploration of objectivity and democracy in group judgment, which builds on my earlier work (2006).

Chapters 5 and 6 are on evidence-based medicine. Chapter 5 describes the origins of evidence-based medicine and shows the continuities between evidence-based medicine and the Greek concept of "empiric"

---

[19] For example, Havi Carel's project of putting together a "phenomenological toolkit" (Carel 2012) and Leah McClimans's (McClimans 2010, McClimans & Browne 2011) project of measuring patient-centered outcomes.

[20] In fact, the NIH Consensus Development Conference Program was the model for all consensus conferences—those on science and technology as well as medicine. I am currently writing a paper, "The Evolution of Consensus Conferences," which traces the history of the institution, from its origins at the NIH to its adaptation into "the Danish Model" and back to the US as a model of participatory democracy.

medicine. Like empiric medicine, evidence-based medicine is not a complete methodology of science and is unlikely to progress without the addition of causal (specifically pathophysiological) reasoning and basic science experimentation. Chapter 6 explores the ways in which evidence-based medicine has turned out to be less reliable than expected, something to be taken into account when making normative claims about evidence-based medicine.

Chapter 7 introduces translational medicine as a predictable reaction to a decade of dominance of evidence-based medicine. It is also a response to the shortfalls, during the 1990s, of promising initiatives from the Human Genome Project and stem cell research, and to the long-term underperformance of consensus conferences in disseminating successful new medical technologies. In these ways, Chapter 7 brings together themes from earlier chapters with observations on recent progress in biomedical research. Chapter 8 is a discussion of the most prominent current example of humanistic medicine and humanities-in-medicine, namely, narrative medicine. The scope of the medical humanities comes into better focus after the previous chapters have explored a variety of scientific and other methodologies in medicine. Finally, Chapter 9 draws on all the previous chapters and looks at how the methods work together, especially when they apply to the same topics. I describe the situation as developing, untidy, methodological pluralism. Disagreement is sometimes "managed away," but on occasions cannot be managed and results in a public medical controversy. I end Chapter 9 with a discussion of the main difference between scientific controversy and medical controversy. Chapter 10 makes a concluding statement.

# 2

# "NIH's Window to the Health Care Community"

## The NIH Consensus Development Conference Program

## 2.1 Introduction

The NIH Consensus Development Conference Program began in 1977 and ended in 2013.[1] It was the model for creating other consensus conference programs in medicine, both nationally and internationally. It conducted over 160 such meetings, producing mostly consensus statements and occasionally, when there was not enough evidence for consensus, State-of-the-Science Statements. (Before 2001, State-of-the-Science Statements were called Technology Assessment Conferences.) Widely commended and imitated in its early years, it became a social epistemic institution with a history of some change and adaptation to criticism and changing circumstances, and some resistance to change. In later years, it tried to accommodate one of its greatest challenges: the rise of evidence-based medicine. It is a rich topic for social epistemic investigation. I will show in this chapter that the NIH Consensus Development

---

[1] My article "The social epistemology of NIH consensus conferences" (Solomon 2007) was a preliminary version of some of the material in this chapter. Just before I wrote the final draft of this chapter, the NIH Consensus Development Conference Program was "retired" (see <http://consensus.nih.gov> accessed October 30, 2014). It was already much diminished from its peak of activity during the late 1980s to late 1990s, so this has not significantly changed my narrative. It has, however, necessitated converting the narrative into the past tense.

Conference Program performed important epistemic functions. Ironically, however, those functions did not include the most frequently stated goal, which was the formation of consensus on controversial topics in medicine.

My account is based mostly on the published and public record, which includes journal articles, NIH publications (printed and online), internal evaluations, and external evaluations of the program. Occasionally I supplement this record with materials taken from my communications with past program directors and their staff.[2] The goal is to reveal the epistemic character of the program, and to show how epistemic tensions developed and were responded to over time. After this discussion, it will be possible to assess the overall role of NIH Consensus Conferences in medicine and to lay the ground for assessment of other medical consensus conferences in Chapters 3 and 4.

There will be considerable detail in my discussions of medical consensus conferences, much more detail than I give in later chapters about other methods. The details are important because they reveal much about the epistemic character of this social epistemic institution. For those whose interests are more historical than philosophical, these details may not be sufficient. Indeed, there is room for more archival and oral historical work than I have been able to do. So far, medical consensus conferences have not attracted the attention of historians or sociologists of medicine. My account is, I hope, a good first approximation of the history of the NIH Consensus Development Conference Program together with a philosophical assessment.

## 2.2  Origins: The Official Story

During the post-World War II years, the United States increased public investment in medical research at the NIH. In the years 1945 to 1975, the number of full time employees at NIH increased from 1,000 to almost 14,000.[3] By the 1970s, concern was expressed in Congress that the new

---

[2] These were informal communications, by e-mail, telephone, and in-person discussion, and I treat them as such: my records may be fallible, and statements were not official communications. I am not trained in social science methodologies, so I do not foreground these informal communications or depend on them crucially in arguments.

[3] In contrast, from 1975 to the present, the number of employees has only risen by 4,500. My source for this information is the NIH website <http://www.nih.gov/about/almanac/staff/index.htm> accessed February 25, 2009 (and no longer available).

medical technologies developed at NIH at taxpayer expense were not being put into appropriate clinical use (Ferguson 1993). The late Senator Edward Kennedy chaired the Senate Subcommittee on Health that in 1976 called for accelerating "technology transfer" between researchers and health care providers (Kalberer 1985). From the beginning, the thinking was that these technologies need both unbiased evaluation and effective dissemination. In response, the then director of the NIH, Donald Fredrickson, created the new OMAR, and instituted the NIH Consensus Development Conference Program.

The official NIH story (Mullan & Jacoby 1985, Jacoby 1988, Ferguson 1997) is that Fredrickson adopted the model of "science court" proposed by Arthur Kantrowitz in two well-known *Science* articles (1967, Kantrowitz et al. 1976). Kantrowitz designed this "institution for scientific judgment" to help resolve situations in which collective action, such as the implementation of a particular technology, needs to be taken before a scientific issue is definitively settled. In such cases, decisions can be complicated by ethical and political considerations. Kantrowitz's aim was to *separate* the scientific and ethical/political considerations with the goal of reaching more *objective* decisions. He used a *judicial* model for epistemic purposes, arguing that an unbiased judge(s), scientifically literate but neutral with respect to the controversy at hand, is suitable for making objective judgments.[4] Kantrowitz imagined that such a "science court" would publish its results and feed them into Congressional decision making, where ethical and political considerations would be added to the (now purified) scientific conclusions. He envisaged a science court as either one judge or a panel of three judges,[5] using scientific standards of evidence, listening to public testimony and discussion from both sides, and then deliberating in private and coming to a conclusion. Kantrowitz had in mind that a "science court" would convene over then controversial matters such as "Should the use of fluorocarbons be banned?" and "Is Red Dye #40 safer than Red Dye #2?" He was under no illusions that a science court would deliver "the truth": instead, the hope was that scientific knowledge could be assessed objectively—without bias—during times of

---

[4] I am not endorsing Kantrowitz's construal of objectivity, just describing it. It is a common traditional account of objectivity (recently challenged in several literatures, including the heuristics and biases literature and the feminist epistemology literature).

[5] Kantrowitz (1967) has one judge and Kantrowitz et al. (1976) has a panel of three judges.

scientific uncertainty. Kantrowitz chaired a Presidential Advisory Group and hoped to implement a general "science court" for governmental use. However, Fredrickson's adaptation for the NIH Consensus Development Conference Program was the closest that Kantrowitz's model came to implementation.

The original idea was that the NIH Consensus Development Conference Program would make the pure science determination and hand off its results to a new policy organization within the Public Health Service, the National Center for Health Care Technology (NCHCT), which would then produce concrete practice recommendations (Mullan & Jacoby 1985, Lowe 1980a). This would parallel Kantrowitz's proposal that the results of science court should be delivered to Congressional committees. The hand off to NCHCT happened only once, after the 1980 conference on coronary bypass surgery, because NCHCT was abolished shortly after it was created[6] (Jacoby 1990). The NIH maintained and continues to maintain that, as a scientific research institute, it is not in a position to make practice or policy recommendations.[7]

Accordingly, since the early days of the NIH Consensus Development Conference Program, "technical consensus" (consensus on scientific matters) has been distinguished from "interface consensus" (consensus on broader societal, ethical, economic, etc. matters) (Fredrickson 1978, Perry & Kalberer 1980). At first, NIH occasionally did both, but quickly settled down by 1982 to doing only technical consensus, despite occasional criticisms of this narrow focus.

Although Kantrowitz's "science court" was an important inspiration for the NIH Consensus Development Conference Program, it was not straightforwardly implemented; it was adapted. There are important differences between Kantrowitz's "science court model" and the NIH

---

[6] I do not know why the Agency for Health Care Policy and Research (AHCPR) (now the AHRQ), which was founded in 1989, or its precursors, did not take up this project. Perhaps the NIH project had already established its role in public deliberations (e.g. by directly informing clinicians) without feeding directly into an organization producing policy and guidelines. And/or perhaps political concerns about the role of government in setting health care policy, which played a role in the demise of NCHCT, discouraged any attempt to produce practice recommendations. I am grateful to John R. Clarke for helping me locate documents about the history of these organizations.

[7] One exception to this is an early paper by Perry and Kalberer (1980), which suggests that the NIH ought to include social, moral, economic, and political implications. The official line since then has been to eschew "non-scientific" implications.

Consensus Development Conference Program. Moreover, there were other models—I have identified eight of them—that played a role in the development of the NIH Consensus Development Conference Program. I will explain these models in Section 2.3. In this section I summarize the epistemically important differences between Kantrowitz's "science court" model and the NIH Consensus Development Conference Program, and describe the basic design of an NIH Consensus Development Conference.

Perhaps the most important difference between Kantrowitz's science court and the NIH Consensus Development Conference Program is that "science court" was not described as a process of coming to consensus but, rather, as a way of making an unbiased decision. Indeed, in his 1967 article Kantrowitz proposes one judge; in the later 1976 article he proposes three judges (no reason for these differences is given). A science court operates on the model of individual judgment (perhaps three judges are intended as checks on one another's normal human fallibility). In contrast, the NIH Consensus Development Conference Program was, from the beginning, conceived as a *panel* decision, with the panel consisting of 10–20 people. The NIH Consensus Development Conference Program is an essentially *social* epistemic practice, and makes use of the epistemic appeal of expert group judgment (I will say more about the epistemic appeal of expert group judgment in Chapter 4). The word "development" signifies that consensus does not exist beforehand, and is to develop during the conference.

Another important difference is that Kantrowitz intended to use a science court for cases where the technology is "so new that debatable extrapolation of hard scientific fact [is] required" (1967) yet a decision needs to be made in order to take necessary action. The NIH Consensus Development Conference Program, on the other hand, was intended to accelerate the usual process of scientific consensus, but not to compromise on the scientific conclusiveness of the decision. When the evidence available was insufficient for producing such a conclusion panelists were encouraged to say so, and perhaps to produce a majority and a minority report or a "State-of-the-Science" report (called a "Technology Assessment Conference" before 2001) reflecting the current uncertainty.

The basic design of an NIH Consensus Conference was settled by 1982, as follows. Topics could be suggested by any governmental body, organization, or individual. The NIH OMAR[8] decided which topics to pursue. A planning committee—which included the preselected panel chair, NIH employees, and outside experts—convened several months in advance to decide on the questions framing the conference. Panelists were chosen from clinicians (mostly physicians but sometimes nurses or other health care professionals), researchers, research methodologists, and patient representatives. Federal employees were not eligible to participate as panel members, so as to avoid the appearance of government influence. Proposed panel members were disqualified if their own research or stated opinions could be used to answer the questions under debate (this is termed "intellectual bias"[9]). Starting in 1993, proposed panel members were also disqualified if they had a financial conflict of interest. The entire cost of the consensus conference was covered by NIH, in order to avoid conflicts of interest with professional or industrial groups. Speakers were requested, but not required, to disclose financial conflicts of interest. Speakers were not expected to be without "intellectual bias," but they were asked to present as researchers rather than as advocates. Members of the planning committee could also be speakers. Members of the panel could never be speakers.

Presentations and the questions following them at NIH Consensus Conferences were always open to the public, and many of those since the year 2000 were made available in video format[10] for remote and later viewing. Panel members and the audience listened to academic presentations from 20–30 experts on the debated issues, with opportunities to ask questions. The ratio of presentation to discussion was about 2:1. In "executive session" (i.e. without speakers, audience, or video cameras present) the panel began to draft a consensus statement on the first evening of

---

[8] This office was abolished in January 2012 (after I wrote the first draft of this chapter) and the NIH Consensus Development Conference Program was then run for a couple of years by the NIH Office of Disease Prevention before being "retired" (after I wrote the second draft of this chapter).

[9] This disqualification often evokes surprise, because those with the greatest expertise on the questions are likely to have stated their assessments already. Those designing the NIH Consensus Development Conference Program decided that those already involved in a controversy do not have the necessary objectivity to be panelists, and that it is preferable to have panelists with less investment in the outcome (even if that means that they will have less expertise). I will say more about this feature of the NIH Consensus Development Conference Program in Sections 2.5 and 3.2.

[10] Now at <http://consensus.nih.gov/historical.htm> accessed October 25, 2014.

the conference, and continued on the afternoon of the second day, often working well into the night. On the morning of the third day, the draft consensus statement was read aloud to all attendees, and comments and discussion welcomed. The panel recessed for about two hours to incorporate any desired changes, and then a press conference was held releasing the major conclusions of the consensus conference. The consensus statement went through one more round of revisions from the panel before final publication about a month later. The consensus statement was released to medical practitioners, policy experts, and the public, originally as brochures and in journal articles, and in later years as PDFs on the program's website.

Topics had to satisfy criteria of appropriateness for a consensus conference. The most important criteria were (1) medical importance, (2) the existence of controversy, i.e. lack of consensus in the research community *or* a gap between current knowledge and practice in the clinical community, and (3) sufficient data to resolve the controversy "on technical grounds" (Institute of Medicine 1985). Timing was therefore important: not so early that the data were insufficient to resolve the controversy, and not so late that consensus was already reached in the profession by the usual processes of dissemination (Institute of Medicine 1985).

NIH Consensus Development Conferences were held on a wide variety of topics in the diagnosis, prevention, and treatment of disease. Some examples are Breast Cancer Screening (1977), Intraocular Lens Implantation (1979), Drugs and Insomnia (1983), Prevention and Treatment of Kidney Stones (1988), Helicobacter Pylori in Peptic Ulcer Disease (1994), Management of Hepatitis C (1997), Celiac Disease (2004), and Hydroxyurea Treatment for Sickle Cell Disease (2008). The frequency of the conferences changed over time, from an erratic early period to a stable period between the mid-1980s and mid-1990s of approximately six conferences a year, to a decline and restabilization at approximately three conferences per year to, on January 12, 2012, an announcement that NIH Consensus Development Conferences are "expected to be held less frequently"[11] and finally to the "retirement" of the program in 2013.[12] Explanations of these changes will be considered later in this chapter.

---

[11]  <http://consensus.nih.gov/CDPNews.htm> accessed June 5, 2012.
[12]  See <http://consensus.nih.gov> accessed May 21, 2014.

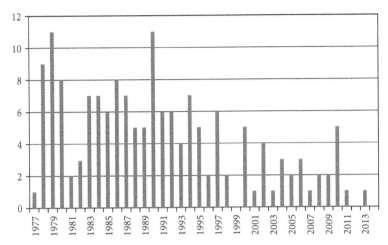

**Figure 2.1** Number of NIH Consensus Development Conferences held annually, from 1977 to 2013

Figure 2.1 presents my tabulation of the number of conferences per year throughout the history of the program.

## 2.3  Origins: The Full Story

I have found no less than nine models and five metaphors for the NIH Consensus Development Conference Program, all present in the literature of the early years of the program. The "science court" already described is just one of the models. Other models are a court of law, the usual scientific process, a scientific meeting, a town meeting, peer review of scientific grant proposals, the US government balance of powers, collective bargaining, and a new biomedical technology. None of these models fits exactly—but that is a general feature of modeling strategies, something to be aware of rather than to judge negatively.[13] In my view, the presence of all these models—all of social institutions rather than of individual skills or reasoning—is evidence of the richness of creative effort and communicative skill that went into crafting and disseminating a new social epistemic technology.

[13] Philosophers and psychologists of science have written extensively about the role of models in creativity and scientific reasoning. See, for example, Boden (1994), Giere (1988).

## I   Court of Law (Mullan & Jacoby 1985)

Since the court of law was a model for Kantrowitz's "science court" as well as directly for the NIH Consensus Development Conference Program, some of the features of this modeling have already been mentioned. The NIH Consensus Development Conference Program took on various features of the US law courts: testimony from all sides, a judge(s) who was expected to be neutral, and proceedings other than private deliberations of judges open to the public. In a 1998 talk by John Ferguson (Director of OMAR during the 1990s), I also heard a specific comparison with the US Supreme Court, in which minority opinions are not ignored, but reported along with the majority statement.[14] (For NIH Consensus Development Conferences, a minority report has been produced only three times.) Some features of US law courts were explicitly disavowed, e.g. Mullan and Jacoby (1985) and Ferguson (1997) point out that there are no winners or losers in a consensus conference, and presentations by experts were expected to model scientific presentations rather than those of adversarial lawyers.

## II   Usual Scientific Process (Fredrickson 1978, Asch & Lowe 1984)

The history of science typically tells narratives of scientific disputes followed by consensus, as Kuhn (1962) and many others have noted. The NIH Consensus Development Conference Program did not lower scientific standards of evidence for the sake of coming to consensus more quickly, and in this way it differed from Kantrowitz's imaginary "science court" which was willing to go with less than conclusive evidence so long as neutrality was preserved. Without lowering standards of evidence, the NIH Consensus Development Conference Program claimed to "hasten resolution of scientific issues" (Fredrickson 1978) as well as to accelerate the transfer of new medical technology (Mullan & Jacoby 1985). Thus it claimed to achieve, within three days, the careful evaluation of reasonably conclusive evidence that is deemed to take place during the typical decentralized, informal, and slower natural processes of coming to

---

[14] The talk was at the Philosophy of Science Association Meetings, October 1998, at a symposium on medical consensus conferences that I organized.

consensus on a scientific matter.[15] Asch and Lowe (1984: 377) explicitly described the NIH Consensus Development Conference Program as modeled on the usual scientific process:

[The NIH Consensus Development Conference Program] is an attempt to reproduce in microcosm, at one time, and in one place, the process of knowledge evaluation, transfer and transformation that ordinarily occurs within the context of the entire biomedical system and its contingent systems . . . Currently, this reproduction of the whole societal process is embodied in a "Consensus Conference" conducted as a public hearing, with predetermined choreography.

## III  Scientific Meeting

During NIH Consensus Development Conferences, speakers were asked to present as though they were at a scientific meeting. In this respect, the model differs from the kind of adversarial model suggested by courts of law in the United States. Speakers are expected to present in as balanced and judicious a manner as possible, just as they would at a professional conference. Speakers are, of course, still not regarded as "neutral." Some intellectual bias is thought of as ineliminable—just as it is at scientific conferences.

## IV  Town Meeting

Town meetings are open to everyone and forums for conversation between the many parties that can contribute to, and may be affected by, a decision. In particular, town meetings are open to "non-experts." From the beginning, the NIH Consensus Development Conference meetings have been open to the public—except during "executive sessions" when the panel meets in private—and there has been time set aside for public questions and comments. Both town meetings and NIH Consensus Development Conferences are seen as socially democratic and politically transparent institutions. However, NIH Consensus Development Conference meetings are much longer and more structured than a town meeting (which typically takes place in the evening after a full working day and dinner). Also, decisions are taken by the panel, in executive sessions at NIH

---

[15] Some historians and sociologists of science, e.g. Collins and Pinch (1993) and Shapin (1994), have argued that consensus in science is more socially constructed than it is decided by the evidence. This will be considered in Chapter 4, with the epistemology of group consensus.

Consensus Development Conference meetings, and not by the audience, or in public, as often happens at a traditional town meeting.

## V  Peer Review of Scientific Grant Proposals

The panel on an NIH Consensus Development Conference can be thought of as like a National Science Foundation (NSF) or NIH review panel. It is composed of experts in fields that are closely related to the material under review, rather than of laypersons.[16] Panelists are screened for conflicts of interest.[17] And just like the typical grant review panel, the meeting is two to three days long and close to Washington, DC. John Kalberer, who served as Deputy Director of OMAR, wrote of NIH Consensus Development Conferences that "This method . . . can be viewed as a modification of the highly regarded NIH peer review system" (Kalberer 1985: 63).

## VI  US Government Balance of Powers

This is hinted at in Mullan and Jacoby (1985) when they write about the "triad of forces" that is responsible for the final consensus product: the planning committee, the speakers, and the panelists. Mullan and Jacoby go on to say that "The separation of authority between these three groups to the greatest extent possible stands as a governing principle of the current process" (1985: 1070). In another publication the same year, Jacoby explicitly uses the phrase "separation of powers" to describe the triad of forces (1985: 428).

## VII  Collective Bargaining

This is mentioned in passing by Jacoby (1990: 8) as a conflict resolution model, but it is not elaborated, so I can only speculate on whether or not it played much of a role. Collective bargaining is typically negotiation and

---

[16] Not all consensus conferences take this form. In particular, European consensus conference programs sometimes have panels entirely composed of public/lay volunteers. Most panelists at NIH Consensus Development Conferences, however, are physicians or researchers in the general area of study, with only the occasional other professional (e.g. nurse, lawyer) or layperson (e.g. patient representative).

[17] At first, intellectual conflicts of interest were taken especially seriously. Other medical consensus conference programs (such as that of the IOM) have been less concerned about intellectual conflicts of interest, seeking a balanced panel rather than a neutral panel. See Chapter 3 for details. As the general concern with financial conflicts of interest grew in the 1980s and 1990s, NIH started to pay attention to this also. In 1993 panelists with financial conflicts of interest were disqualified from participation in NIH Consensus Development Conferences.

compromise between adversaries, a process to be settled through creation of a mutually acceptable choice. It does not model the "evidence-based" goals of NIH Consensus Development Conferences, but perhaps it captures the experience that, in practice, some creativity is needed to draw conclusions that all are willing to accept.

## VIII   New Biomedical Technology

The NIH Consensus Development Conference Program was thought of as analogous to new biomedical technologies, as something to be "disseminated, adopted and adapted" (Rogers et al. 1982: 1882). In fact this happened, as the NIH Consensus Development Conference Program became a model for both national and international consensus conferences in medicine and other fields. The idea was also that the NIH Consensus Development Conference Program, like a biomedical technology, could be evaluated for its effectiveness. Measures of effectiveness would of course have to be specified. (This is a somewhat reflexive analogy, since the NIH Consensus Development Conference Program was itself designed to evaluate new biomedical technologies. So now we have a new biomedical technology for the evaluation of biomedical technologies in general.)

In addition to these eight analogical models, I have noticed five metaphors for the NIH Consensus Development Conference Program in the literature (I expect that there are more):

1. Asch and Lowe claim (in a rather mixed metaphor) that the purpose of the NIH Consensus Development Conference Program is to relieve the "bottleneck" in the "knowledge explosion" (1984: 371).
2. Asch and Lowe also use the metaphor of "a beacon, a lighthouse, upon which both practitioners and consumers can take bearings" (1984: 383). Perhaps I can elaborate that metaphor as an "epistemic lighthouse."
3. Seymour Perry (1988: 482) describes the role of the NIH Consensus Development Conference as that of a "catalyst" to help bring about a "technical consensus" (by which is meant a consensus on scientific matters). This goes together with model III above: modeling the NIH Consensus Development Conference as an acceleration of the usual scientific process.
4. John Ferguson, one of the longest serving directors of OMAR, said that "Consensus conferences are intended to serve as NIH's

window to the health care community" (1995: 333). This indicates that NIH Consensus Development Conferences are, not surprisingly, especially concerned to focus on research done at NIH or with NIH funding. Also, the metaphor reminds us that NIH controls—via its selection of topics—where the window is placed. I find this to be a particularly informative metaphor, and use it for the title of this chapter.

5. Most recently (since 2000), the consensus conferences have been described on their official NIH website as a "snapshot in time" of the state of knowledge on the conference topic.[18] Conference statements more than five years old have been moved to an archive and "provided solely for historical purposes."[19] This new metaphor of a "snapshot in time" may reflect increased awareness of the fallibility of both expert consensus and evidence-based medicine. It also sends a message that up-to-date evidence is essential.

These analogical models and metaphors helped to shape and to communicate the epistemic characteristics of the NIH Consensus Development Conference Program. Recent literature on the psychology of creativity suggests that analogical and metaphorical thinking is indispensable in the creative process, as well as helpful in communication and pedagogy (see, for example, Boden 1994, 2004). I think the presence of so many models and metaphors—none of which fit perfectly, but all of which fit in some important way and for some time—are evidence of the creative effort that has gone into creating a new social epistemic technique, and the rhetorical effort that has gone into communicating it to the intended audience.

One general limitation of these models and metaphors is that they are culturally constrained by the ideals and norms of the culture of the United States. Some social institutions that NIH Consensus Development Conferences are modeled on are the United States' legal system and the United States' government balance of powers. It is possible that the institution of NIH Consensus Development Conferences is limited in ways that are difficult for Americans to imagine. It will be helpful to take a look at the institutions of medical consensus conferences in other countries (see Chapter 3).

---

[18] <http://consensus.nih.gov/aboutcdp.htm> accessed June 6, 2012 (and no longer available).

[19] <http://consensus.nih.gov/historical.htm> accessed June 6, 2012.

## 2.4  Early Assessment and Criticisms (1978–88)

In general, the NIH Consensus Development Conference Program was greeted with enthusiasm. Phil Gunby, a *Journal of the American Medical Association* editor and early critic of the program, wrote that officials of the program reported that "The mail is running about 400:1 in favor of the program" (1980: 1413). Participants in the conferences were generally very positive (Wortman et al. 1988, Ferguson & Sherman 2001). Nevertheless, true to the model of "a new biomedical technology," implementation was viewed as experimental, and assessment was built into the program from the beginning. There was outside criticism, formal outside evaluation, and inside criticism. In this section I will consider the first decade of the program (1978–88).

Concern about the epistemic quality of the NIH Consensus Development Conference Program includes concerns both about the epistemic value of consensus generally speaking (the scientific community's consensus has often been wrong, as with the motion of the earth, the practice of alchemy, belief in miasmas, etc.), and about the *particular* strengths and weaknesses of the *particular* process that NIH used to come to consensus (i.e. the details of the deliberative process over two-and-a-half days). After some introductory remarks on the general epistemics of consensus, I postpone lengthier discussion of that topic to Chapter 4. Then the remainder of this section focuses on the *particular* strengths and weakness of the NIH Consensus Development Conference Program during its first decade.

From the beginning, a few individuals raised general skeptical questions about the rationality of the process of coming to consensus. Gunby, in the *Journal of the American Medical Association* editorial mentioned at the beginning of this section, reported Alfred E. Harper's statement that "Scientific controversies can be resolved only by rigorous testing and critical evaluation . . . Scientific controversies, unlike public policy decisions, cannot be resolved by consensus" (1980: 1414). Gunby also expressed the concern that it will be difficult for individual physicians to challenge the authority of consensus statements, which is unfortunate for those consensus statements that are "inappropriately arrived at" (1980: 1413). More bluntly, Arthur Holleb (editor in chief of *CA: A Cancer Journal for Physicians*) wrote an editorial deploring the idea of "medicine by committee" and argued that the potential for coming to a biased conclusion is inescapable (1980). Holleb disapproved of the "media oriented nature" of the NIH Consensus Development Conference Program and feared that the resulting guidelines would be used for reimbursement decisions and medical malpractice cases. Holleb was also concerned that the guidelines would be too broad to include the nuanced judgments needed for the care of individual patients. This point is taken up in a letter responding positively to Holleb's article from the Director of the American College of Surgeons, C. Rollins Hanlon (1981), which characterizes the unwelcome result of consensus conferences as "cookbook medicine," a derisory term later also applied to evidence-based medicine.[20] Drummond Rennie, who became a leader in evidence-based medicine, wrote "I fear lest these statements will act like a dead hand, discouraging thought" and insisted that expert opinion should not replace clinical trials (1981: 666). Rennie also feared that the most that could be expected from the NIH Consensus Development Conference Program was "the bland generalities that represent the lowest common denominator of a debate" (1981: 666).

Gunby, Harper, Holleb, and Rennie were exceptions to the typical enthusiastic reactions to the early years of the NIH Consensus Development Conference Program. But their concerns deserve attention. Typically, when scientists disagree they do *not* convene a consensus conference in order to resolve their disagreements. They continue their research and often provide critiques of each other's work. Coming to consensus does, however, play an important role for some kinds of social decision making. The idea that a group of qualified persons can rationally deliberate a controversial issue to the point of rational consensus when the need arises is powerful both politically and intellectually. Social institutions and practices that are regulated by such procedures and ideals include governing boards of public and private institutions, federal investigative and advisory panels, and grant study

---

[20] The term "cookbook medicine" is typically used to denote "one size fits all" or "medicine not tailored to individual patients." It is seen not only as inferior medicine but as a challenge to physician autonomy and patient individuality. Its role in debates about medical practice is complex, and will be discussed in later chapters, especially Chapters 5, 6, and 8.

sections. The idea of John Rawls's "original position" includes rational deliberation by groups (behind the veil of ignorance) as the foundation of assessments of justice. More prosaically, the endless series of committee meetings that many of us suffer through are designed with the same ideal of rational group deliberation. Their purpose is not simply political—to ensure the appearance or reality of democracy (for that, a mail ballot would be enough) or to forge *any* agreement for the purposes of action—but also epistemic, to ensure the best (most effective, or most "rational") decision by permitting members of the group to improve their positions by listening to the ideas and criticisms of other members of the group. Scientists also engage in mutual criticism and debate, but they do not expect to use it to resolve all their differences. When scientists continue to disagree they go back to the bench or the field, and the notebook or the computer, to develop their ideas and test them further empirically. Lack of consensus is often productive in science, as I argued in Solomon (2001). On the rare occasions when scientists attempt to come to consensus—such as for the statements of the Intergovernmental Panel on Climate Change (IPCC)—it is not because they have important differences with one another that they want to resolve, but because they want to take a united (and therefore strong) stand in scientific matters of public concern. Such expert consensus is rhetorically persuasive, making use of the perceived authority of the consensus panel to persuade non-experts to accept its deliberations on trust. I will say much more about these matters in Chapters 3 and 4.

I return now to the specific assessments of the NIH Consensus Development Conference Program in its first decade. From the beginning, there was concern to structure the consensus conferences so that the process would be, and be perceived to be, as *unbiased* as possible. The NIH looked for "objective" or "neutral" (without "intellectual bias") panelists, and they disqualified federal employees in order to avoid the appearance of an official government agenda. Most proceedings (all except for panel-only deliberations) were open to the public. Nevertheless, there were some epistemic concerns about the process, which emerged over time in both internal and external evaluations, and especially for conferences in which there was controversy: e.g. Treatment of Primary Breast Cancer (1979), Liver Transplantation (1983), and Cholesterol and Heart Disease (1984).

In its early years, the NIH Consensus Development Conference Program was formally evaluated by several external organizations. There was a University of Michigan process evaluation study in 1980-2, two Rand Corporation studies—one focusing on content analysis (1984) and the other focusing on reception of NIH Consensus Development Conferences (1986), and an IOM report in 1985. The University of Michigan study, which used questionnaire feedback from panel participants, found that strong disagreements harmed the perceived quality of the conference. Perhaps in response to this, "questions that dealt with the controversial aspects of a technology have often been eliminated" (Wortman et al. 1988: 477), surely an undesirable result given the stated goals of the NIH Consensus Development Conference Program to provide "a forum in which panel members have been able to reach consensus on a medical technology where no consensus has previously existed" (Lowe 1980b: 1184). The University of Michigan study also raised concerns with the objectivity of the process, reporting that panelists sometimes felt overwhelmed by information and that panelists often complained about the time pressure, and especially about the late-night sessions typically required to develop a consensus in three days. In addition, there was concern about selection bias in the questions chosen for consideration and the experts chosen as panelists. Panelists and speakers often changed their views about which questions were important in the course of the conference. Finally, there was concern in the early years of the NIH Consensus Development Conference Program that panelists were mainly drawn from the "old boy" networks around elite East Coast universities (Wortman et al. 1988).

The Rand Corporation reports observed that most, if not all, NIH Consensus Development Conferences concerned technologies already in mainstream use, rather than technologies still only at the stage of clinical trials (Kahan et al. 1984). So NIH Consensus Development Conferences were not being convened as early as intended, which was after the research is done but before the technology is adopted in the wider clinical context. The first Rand Corporation study also looked at the content of 24 conferences, classifying each sentence in the consensus statement in terms of categories such as "discursive" (few recommendations, rather abstract), "didactic" (practical recommendations), and "scholarly" (assess current research and make research recommendations). The study revealed a "gradual shift to greater discursiveness" (Kahan et al. 1988: 300) which is undesirable in a publication aimed at clinicians and the public. In response, OMAR encouraged panelists to "recommend concrete, specific actions" (Kahan et al. 1988: 303). This encouragement is in tension with the frequently stated desire (and the political necessity) for the NIH Consensus Development Conference Program not to be perceived as dictating the practice of medicine.

The 1985 IOM Report notes that although the official policy is that "conference questions are limited to issues for which sufficient data exist for reaching scientifically valid findings" (Institute of Medicine 1985: 133), on some occasions this policy was disregarded, e.g. for the 1984 conference on dietary cholesterol. This conference recommended lowering dietary cholesterol for all those over the age of two, a recommendation that was supported at best by "suggestive" evidence (Institute of Medicine 1985: 133), and opposed by a number of critics (Kolata 1985, Ahrens 1985, Oliver 1985). The IOM report acknowledges that there is a "conflict between the intent to avoid conference topics for which insufficient data are available for reaching scientifically valid conclusions and pressure to hold conferences on controversial issues" (Institute of Medicine 1985: 133). I will say more about this perceived conflict, as well as other tensions, in Section 2.5.

The 1985 IOM Report expresses concern about the group process itself, noting that "the Consensus Development Program conference format is not designed to limit problems associated with face-to-face interaction (e.g. relative dominance of viewpoints due to social or hierarchical factors) in group settings, as are Delphi, nominal group, and other group approaches" (Institute of Medicine 1985: 394). In fact, the group process during an NIH Consensus Development Conference is quite unstructured and thus vulnerable to biasing social process such as groupthink.[21]

In contrast to the early external reviews, which are somewhat critical, early self-evaluations show excitement about the promise of the NIH Consensus Development Conference Program. A typical example is the somewhat premature enthusiasm and optimism expressed about cross-cultural adaptability of the program after just one Swedish

---

[21] Janis's (1972) work on groupthink was first published in early 1972, so the phenomenon was already known.

Consensus Conference that followed the NIH model. Delight was expressed at finding the two consensus statements "remarkably similar" (Rogers et al. 1982), despite the fact that the Swedish conference was run by staff trained at NIH about two months after the NIH Consensus Development Conference on the same topic.

Early self-criticism focused on concerns about dissemination of results. Itzhak Jacoby, who was involved with the NIH Consensus Development Conference Program for some years, reported on an early study to measure physician awareness of consensus conferences and their results (1983). Less than a third of physicians were aware that relevant conferences had taken place and even fewer were aware of their particular recommendations; Jacoby's conclusion was that "there is significant room for improvement in conveying the consensus message" (1983: 258). Others showed that even when the target audience was reached, conference results "mostly failed to stimulate change in physician practice" (Kosecoff et al. 1987: 2708). Changing physician practice turns out to be a long and difficult process, requiring both knowledge dissemination and incorporation of the knowledge into regular practice. This was not fully appreciated at the time, when there was surprise and disappointment at the lack of uptake of consensus knowledge, despite increased efforts to disseminate knowledge through, for example, direct mailing to physicians. We are still struggling with the difficulties of changing physician practice.

By 1987, almost ten years after the first NIH Consensus Development Conference, concern was expressed within the program about the ways in which panels handled complex data. Although panelists received "a book of abstracts of conference papers and background information" (Mullan & Jacoby 1985: 1071) well in advance, panelists did not always make decisions based on the whole body of available data and were sometimes criticized for being inadequately or partially informed. Both Seymour Perry and Itzhak Jacoby (who served as full or acting program directors in OMAR) called for more systematic information to be provided to the panel, with a "data synthesis" (Jacoby 1988) and "a comprehensive synthesis and analysis of the available literature . . . subjected to peer review before the conference" (Perry 1987: 487). These recommendations, which were not followed at the time, were prescient, emerging as they did before the birth of evidence-based medicine. In fact, the NIH Consensus Development Conference Program did not fully address the concern until 2001, when it partnered with AHRQ to provide a systematic evidence review to the panel about a month in advance of each meeting (see Section 2.6).

Despite these problems, which were raised by outside and inside evaluators from the beginning, the NIH Consensus Development Conference Program had rapid and early success in that it was widely adopted as a model for other medical consensus conferences. By 1988 the University of Michigan study reported that "The Consensus Development Program is arguably the most visible and influential medical technology assessment activity in the United States. It is a technology in its own right and one that has been widely embraced by other nations" (Wortman et al. 1988: 495). This early enthusiasm and widespread adoption in the absence of robust outcome evaluations needs explanation. There may be a need(s) that the NIH Consensus Development Conference Program addressed, despite the mediocre outcome evaluations for its official goals. I will make some suggestions about this in Section 2.7.

## 2.5  The Second Decade (1989–99)

From the late 1980s until the late 1990s, the NIH Consensus Development Conference Program was in its prime. John Ferguson was the director from 1988 until 1999, giving the program a long period of stable leadership. These years were the "golden years" of consensus conferences in medicine, before evidence-based medicine took the lead. Other US consensus conference programs include those of the IOM (consensus conferences begun in 1982), the Medicare Coverage Advisory Committee (founded 1998, renamed in 2007 as the Medicare Evidence Development Coverage Advisory Committee), the American College of Physicians' Clinical Efficacy Assessment Project (founded 1981), and the Blue Cross and Blue Shield Technology Evaluation Center (founded 1985). A 1990 IOM report assessing consensus conferences internationally begins: "Group judgment methods are perhaps the most widely used means of assessment of medical technologies in many countries" (Institute of Medicine 1990b: 1). The NIH Consensus Development Conference Program fell into a regular schedule of five or six conferences per year, and there were few adjustments in the protocol. The most important formal assessment during this period was the 1990 IOM report (Institute of Medicine 1990a), which will be discussed in this section.

The stability of the institution of NIH Consensus Development Conferences did not make the epistemic questions about them disappear. I think that the stability resulted from a balance between competing epistemic demands and, sometimes, from the power of an established institution to ignore certain questions and recommendations. The competing epistemic demands are reflected in the ongoing epistemic tensions of NIH Consensus Development Conferences, which continued through the second decade:

(a) Panelists should have expertise in the topic of the conference; however, they should not have the "intellectual bias" that comes from working in the field as a contributor. So they should know something, but not too much. Mullan and Jacoby (1985: 1071) remark on the need for "balance" here.

(b) Consensus conferences should not be planned so early that there is an insufficient evidence base for rational decisions, yet not so late that there is no longer controversy about the medical intervention. In practice, most consensus conferences took place on the late side (Kosecoff et al. 1987, Riesenberg 1987). This was acknowledged by Program officers (Ferguson 1993) who, in response, emphasized the dissemination (rather than resolution of controversy) role of consensus conferences. A few consensus conferences took place too early, for example the 1983 conference on liver transplantation and the 1984 conference on cholesterol; in both these cases the results of the conferences were criticized in the media and the scientific literature and failed to bring about a wider consensus. The 1985 IOM study (discussed in Section 2.4) acknowledged a "conflict" between pressure to hold conferences on controversial matters and the requirement to choose topics for which sufficient data are available for closure (Institute of Medicine 1985: 133). This tension continued into the 1990s.

(c) There should be enough time in the conference to consider all matters thoroughly, yet not so much time that busy panelists will be reluctant to participate or media attention will fade. There were regular complaints about time pressure (e.g. Oliver 1985), and regular defenses of the need for such an intense process for producing consensus (Mullan & Jacoby 1985). No changes were made in the schedule of the conference.

(d) Consensus statements should be specific enough to avoid being vacuous or bland, but not so specific that they look like they are prescribing the practice of medicine. This tension was discussed in Section 2.4.

(e) Topics selected should be those for which there is scientific controversy, yet discussion of ethical, political, and economic issues is to be avoided; the NIH only speaks to the science. In practice, however, the causes of a particular controversy are not always known in advance. Thus when taking on a controversial topic, the NIH Consensus Development Conference Program did not know in advance whether or not the science was actually controversial, or whether it had been made to seem controversial because of the other issues at stake. Many have found the focus of the NIH Consensus Development Conference Program on "technical" (scientific) consensus to be too narrow for practical use, and in practice NIH Consensus Development Conferences have occasionally addressed social, economic, and ethical matters ("interface" consensus).[22] The 1990 IOM report advised broadening the scope of the conferences to include "relevant economic, social and ethical aspects of assessing biomedical technologies" (Institute of Medicine 1990a: 1) but this advice was not taken.

(f) Decisions should be checked by independent expert review; but such review cannot be endless and should not be seen as driving the process. There were repeated requests for additional levels of review, such as peer review of the questions to be considered at the conference and peer review of the final consensus statement (Institute of Medicine 1990a). Neither of these levels of review was implemented. The NIH Consensus Development Conference Program wanted to preserve the media event at the end of the conference, so external peer review of the final consensus statement, which takes time, could not be added. The 1990 IOM Review also recommended the establishment of an external advisory council to oversee the

---

[22] The difference between "technical" and "interface" consensus is explained in Section 2.2. It originated in Kantrowitz's separation of the scientific and the ethical/policy/economic issues, which in turn reflects the philosophy of science of that time.

NIH Consensus Development Conference Program (Institute of Medicine 1990a). This did not happen.

(g) There is a tension between the desire to keep the NIH Consensus Development Conferences open to the public, which shows them to be open and democratic, and the need for private planning sessions in advance of the conferences and private executive sessions in which to do the business of coming to consensus. For many people, the prestige of a committee is enhanced by the "executive sessions" that take place in private, perhaps in part because of the air of exclusiveness and in part because of the benefits of not seeing sausage made. Harry Marks (1997: 226) has also argued (in another context) that it is important to "negotiate compromises in private," because "Concessions are easier to make when they are not acknowledged as such." So openness and democracy are in tension with prestige and privacy. I am surprised that no one else has commented on this privacy at the core of the NIH consensus process.

Epistemic tensions are normal,[23] and characteristically can never be perfectly resolved.[24] There is no algorithm to follow to balance such tensions. In "Objectivity, Value Judgment and Theory Choice," Thomas Kuhn (1977) described this pluralism and imprecision of epistemic virtues as unavoidable. The best that can be hoped for, in my opinion, is a good balance of complaints and praise from both sides of the various tensions—evidence of what I called "epistemic equipoise."[25]

---

[23] By "normal" I mean that they are pervasive in social epistemic practices and not in themselves epistemically objectionable. The presence of epistemic tensions, with the vagueness of the recommendation for "balance," does suggest that formal algorithms for how to proceed may not be forthcoming.

[24] Perhaps the careful balancing of opposed values is evidence of an engaged epistemic practice. It is certainly familiar from other epistemic contexts, for example Quine's balancing of simplicity and conservatism in talking about scientific theorizing. The sociologist Kai Erikson (in *Everything in its path*, a 1976 book on an Appalachian community after a devastating flood) uses the terminology "axes of variation" to describe the influence of opposed pairs of ideals on a society, observing that we never observe one extreme without the other, and we see shifts along the axes in response to challenges. For example, masculine ideals of independence that are strong in the Appalachian community occur together with dependence on the community. It is a good theoretical model for what I see here. Moreover, the particular balance of virtues that forms in the NIH Program is shifted in other consensus conference programs.

[25] See Section 1.4.

Along with these continuing epistemic tensions, there were four repeated criticisms of the program and suggestions for improvement that were not responded to in the 1990s:

(i) Recommendations were made to provide the panelists, in advance, with a background report systematically reviewing relevant research (Jacoby 1988, Institute of Medicine 1990a). Left alone, panelists were sometimes overwhelmed by the quantity and complexity of information provided, and ended up basing their decisions on a few salient studies. The biased nature of these decisions was sometimes noticed and criticized (Ahrens 1985). This demand for systematic evidence review grew stronger as evidence-based medicine entered the scene in the 1990s.

(ii) Criticisms of the informal group process in the NIH Consensus Development Conference Program led to recommendations for a more structured group process that would avoid anchoring, groupthink, and other biases (Institute of Medicine 1990a, Fink et al. 1984). One suggestion (not taken) was to use a professional facilitator rather than identify a panel chair (Institute of Medicine 1990a). Other suggestions were to use a more formal group process, such as Delphi and nominal group methods. These suggestions were also not taken. I conjecture that the idea of constraining the supposed "free flow" of rational deliberation by structuring the discussion was judged to be undesirable because it could diminish the credibility of the process to a general audience which does not have sophistication about group processes. (More will be said about this in Chapter 3, when there will be examples of more formally structured consensus conferences aimed at specialist rather than generalist audiences.)

(iii) Evaluations of the impact of NIH Consensus Development Conferences continued to find that they had little, if any, effect on practice (Kosecoff et al. 1987, Kosecoff et al. 1990). This did not dampen the enthusiasm for holding these conferences.[26] Ann Lennarson Greer has argued that the culture of practice is appropriately cautious about making changes (1987) but most evaluators

[26] It is still difficult to change medical practice and we have still not lost our enthusiasm for trying, nor our use of group consensus methods in the effort to do so.

continue to think that practitioners are best thought of as "slow scientists" and that they should strive to improve dissemination and uptake.[27]

(iv) The 1990 IOM report strongly recommended that the NIH Consensus Development Conference Program broaden its focus "to include relevant economic, social and ethical aspects of assessing biomedical technologies" (Institute of Medicine 1990a: 1). The authors of the report thought, plausibly enough, that changing the behavior of clinicians requires taking on not only the scientific issues but also the broader social context. They urged that if the NIH Consensus Development Conference Program could not do this, they should collaborate with another organization, such as Agency for Health Care Policy and Research, to do so (Institute of Medicine 1990a). This recommendation was not taken, perhaps because it would conflict with the core "science court" idea of separating the scientific and non-scientific issues, or perhaps because the NIH felt it did not have the expertise or authority to take on wider questions.

## 2.6   Changes from 1999 to the Present

The NIH Consensus Development Conference Program was evaluated by an NIH Working Group in 1999 (Leshner et al. 1999). By this time, evidence-based medicine was well established as a new method for assessing medical technologies, and was often contrasted favorably with group methods and with expert judgment. In fact, group methods and expert judgment are generally classified by evidence-based medicine as low on the hierarchy of evidence quality (if classified as evidence at all). At this point, the often-repeated suggestion that the NIH Consensus Development Conference Program do some formal assessment of the evidence before the panel meeting became urgent and critical to the survival of the program. To quote from the report:

The current approach dates back to the 1970s and the field of systematic reviews and guideline development has progressed far since that time, but has not been incorporated into the Consensus Development Conference process. Other

---

[27] See, for example, Gawande (2013).

models, while taking additional time and resources, may be more appropriate . . . The workgroup felt strongly that the consensus development process itself would benefit from the application of new methods to systematically review data prior to a consensus conference, such as those used in the evidence-based approach used to establish practice guidelines in recent years . . . In this model the OMAR would commission a systematic review of the scientific evidence on a topic as part of the expanded preconference preparation activities. (Leshner et al. 1999: 8)

The Working Group emphasized the importance of "scientific rigor" in the NIH Consensus Development Conference Program, and recommended that consensus conferences not take place "where the science does not warrant" unless the goal of the conference is to point out that the evidence for a particular intervention is inconclusive (Leshner et al. 1999: 4–5). The NIH Consensus Development Conference Program always avoided pushing for consensus that was not adequately evidence based, and this resolve only became stronger in the age of evidence-based medicine. The Working Group also affirmed that the role of the NIH Consensus Development Conference Program was not to make policy recommendations or come up with specific practice guidelines, but rather "to target the moment when the science is there but the practice is not" (Leshner et al. 1999). This affirms the dissemination role of the NIH Consensus Development Conference Program. Finally, the Working Group supported the mission of NIH as a scientific research organization detached from economic, social, legal, and ethical aspects of issues, suggesting that "these should not be the primary focus of what is essentially, and should remain, a scientific evaluation" (Leshner et al. 1999: 5).[28]

In response to the report of the Working Group, the NIH Consensus Development Conference Program entered into a partnership with the AHRQ, which commissions a systematic review of the evidence and gives it to the panel in a one-day executive meeting about a month prior to the consensus conference. This allows opportunity for reflection and follow-up before the conference.

The Working Group also recommended the formation of an advisory group to "review and approve topics, panel membership and

---

[28] This is the same NIH that set aside substantial funding for research on the ethical, legal, and social implications of the Human Genome Project from 1990 on. There are two matters of concern here: Should the NIH engage in discussion of "non-scientific" issues such as social, legal, ethical, and economic issues? And can the scientific issues be separated from the non-scientific issues?

planned speakers to assure balance and objectivity" (Leshner et al. 1999: 1). Note that this was proposed before, in the 1990 IOM report and earlier reports. However, OMAR did not form such an advisory group, perhaps because it would add one more layer of bureaucracy to a process that already had oversight from a planning committee, OMAR, and open public participation. The NIH Consensus Development Conference Program also resisted repeated recommendations to relieve the time pressure of a two-and-a-half day conference which often ran into late-night sessions, "characterized as unduly grueling by many participants, at times resulting in less thoughtfully developed conclusions" (Leshner et al. 1999: 82) The pressure could have been relieved by lengthening the conference, adding additional meetings, or adding a level of serious review before the press conference. The Working Group recommended both preconference meetings and six to eight weeks of thoughtful review after the conference, but these were not introduced. Perhaps the program did not want to give up the motivating force and rhetorical power of a press conference at the conclusion of the meeting. Moreover, the conferences, while perhaps not long enough to get the job done optimally, were probably as long as busy professionals could tolerate.

Producing systematic evidence reports requires bibliographic, statistical, and domain expertise. In the United States, twelve[29] "Evidence-Based Practice Centers" were established in 1997 and funded by the AHRQ. They provide evidence reports that are widely used, particularly by government organizations. From 2001 to 2013, the NIH Consensus Development Conference Program commissioned such evidence reports, distributing them to the panel in advance of the meeting. After a lag around the year 2000, the NIH Consensus Development Conference Program resumed a regular schedule of conferences in the 2000s, although not as frequent as the peak years in the mid-1980s to mid-1990s. Perhaps the decreased frequency was the consequence of a more complicated and time-consuming process, or perhaps it was the result of less perceived need for consensus development conferences in an age of evidence-based medicine. I will say more about this in Section 2.7.

---

[29] As of May 2014 there are 11 evidence-based practice centers. See <http://www.ahrq.gov/research/findings/evidence-based-reports/overview/index.html> accessed May 29, 2014.

The arrangement during the 2000s between NIH and AHRQ only reinforces my sense that consensus development conferences are not really devoted to the resolution of scientific controversy. Since 2001, when an NIH Consensus Development Conference panel commissioned an evidence report on the topic of the consensus conference, they received a report in advance of the conference on the best evidence for the questions or conclusions at hand. So what was left for the panel to do? Why didn't they simply publish the evidence report? How can a consensus conference represent a "scientific evaluation" when that is already accomplished by the evidence report? If more epistemic work is done after the evidence report, isn't that going beyond the evidence? (I will answer these questions in Section 2.7; in this section I am getting all the questions on the table.) It's important here not to equivocate on "evidence." Evidence-based medicine is just one, powerful set of methods of systematically assessing and combining the evidence. So it is *possible* that there is more scientific evaluation for the panel to do, perhaps in the form of evidence that evidence-based medicine does not accept or in the form of a different overall aggregation of the evidence.

Are there any strictly epistemic—information processing—functions that consensus conferences can perform in an age of evidence-based medicine? After all (as I will argue in Chapter 5) evidence-based medicine is not a complete epistemology of medicine. It does not, for example, give us guidance with scientific theorizing and causal reasoning. Evidence-based medicine produces epidemiological results about correlations rather than scientifically plausible accounts of causal mechanisms. This latter kind of theorizing, however, is typically done by research scientists and not by a group consensus process.

In telephone communication with me in 2002, Susan Rossi (then Deputy Director of OMAR) maintained that consensus conferences do epistemic work—specifically, the work of "extrapolating" experimental studies to ordinary clinical contexts and "comparing risks and benefits."[30] For example, in recommending against universal screening for celiac disease, the 2004 Consensus Statement on Celiac Disease took into account the base rate of the disease in the US population, and a comparison of risks (unnecessary surgery and unnecessary diet for those

---

[30] The telephone conversation was on September 23, 2002.

testing positive to antibodies but without organic disease) and benefits (avoidance of malabsorption syndromes and bowel lymphomas).[31] Application to local circumstances and comparison of risks and benefits are both necessary, but are they tasks that are best accomplished by an expert consensus conference? Wouldn't a formal evaluation using the techniques of medical decision making and estimates of utilities from patients be a more appropriate method? In any case, wasn't this a departure from the stated mission of the NIH Consensus Development Conference Program, which was to answer the scientific questions only and not to balance benefits and risks? The addition of a systematic evidence review to the NIH Consensus Development Conference Program only reinforces the sense that it does not (and should not) hand off any significant information processing to the panel. If NIH Consensus Development Conferences do not function to resolve scientific controversy, why did they continue after meta-analysis and systematic evidence review (from evidence-based medicine) took over the business of aggregating complex evidence?

I suggested at the end of Section 2.4 that NIH Consensus Development Conferences satisfied some need(s), even though they generally did not succeed in resolving controversy or disseminate practice. They satisfied this need(s) not because of the outcome of the deliberations (these were practically guaranteed by the AHRQ evidence report) but because of the process that was followed, which was designed to satisfy demands for "objectivity." In these days of evidence-based medicine, at least some of the audience for NIH Consensus Development Conferences is methodologically sophisticated enough to demand that conclusions be evidence-based. But some place for group judgment, especially the judgment of experts in the prestige of an NIH setting, continues. Evidence-based medicine has not replaced group judgment as our standard of objectivity but, rather, supplemented it. In Section 2.7 I will develop some suggestions about the actual epistemic role of NIH Consensus Development Conferences.

Some of the resistance to change in the NIH Consensus Development Conference Program and some of the reason for the continuation of the program was probably institutional inertia. The NIH Consensus

---

[31] The balancing of risks and benefits was made informally. Perhaps formal methods would be an improvement.

Development Conference Program was so well received in its early years, and so influential in the creation of other consensus programs, that it acquired a status and a reputation that carried it forward for a long time. It did not have to respond to *every* criticism in order to survive. It did, however, need to respond to the rise of evidence-based medicine in order to survive into the twenty-first century. Its response, although satisfactory in some ways, was not sufficient to keep the program alive for much longer.[32]

On January 10, 2012, the NIH announced the dissolution of OMAR, which was created to run the Consensus Development Conference Program, and absorbed its activities into the Office of Disease Prevention. It stated that NIH Consensus Conferences "are expected to be held less frequently."[33] The first—and, as it turned out, the last—conference under the Office of Disease Prevention was held in March 2013. Over the remainder of 2013, the Office of Disease Prevention was engaged in mission planning. In early 2014, the following announcement appeared on the NIH Consensus Development Conference's webpage:

In 2013, the Office of Disease Prevention (ODP) formally retired the Consensus Development Program ... the NIH CDP has served as a model for consensus conference programs developed in many countries ... The CDP was created during a time when few other organizations were providing evidence reviews. Today, there are many organizations, that conduct such reviews, including other federal agencies, academic institutions and private organizations. Examples include the U.S. Preventive Services Task Force, The Community Preventive Services Task Force, the Institute of Medicine, and The Cochrane Collaboration. The CDP has served a very useful role, but one that is now served by other able parties.[34]

There was no wider announcement of the "retirement" of the NIH Consensus Development Conference Program. I think that this is a rather sad fade out: this minimal announcement does not even call the program by its given name (it omits the word "conference"), it confuses consensus conferences with evidence-based medicine, and the entire statement was "off the radar" in that, so far as I know, it was not noticed and commented on in the medical or the wider community. In May 2014 I had

---

[32]  I do not mean to imply that the program should have been kept alive longer. I am trying not to make any judgments until the end of the chapter.

[33]  <http://consensus.nih.gov/CDPNews.htm> accessed February 13, 2012 (and no longer available).

[34]  <http://consensus.nih.gov> accessed October 30, 2014.

a brief e-mail correspondence with David Murray, the current Director of the Office of Disease Prevention, who confirmed the reasons for the decision to discontinue the NIH Consensus Development Conference Program and added "the mission of the NIH is the *development* of new knowledge . . . if we focus on that mission, the[n] many other able groups can continue to summarize the evidence and draft guidelines for practitioners, organizations and communities."[35] This statement overlooks the dissemination role of the NIH Consensus Development Conference Program that has been important since its early days. However, it does not give up the important role of group consensus; it just shifts it to "other able groups." The use of the word "retirement"—rather than, say, the word "discontinuation"—signifies that the work is valuable and continues elsewhere.

## 2.7  Analysis

The NIH Consensus Development Conference Program, despite its name and one of its stated goals, did not function to develop a "technical" consensus on controversial scientific matters. (It did not develop "interface" consensus either, since it eschewed political, economic, ethical, and social matters.[36]) Despite the claims of its supporters, NIH Consensus Development Conferences did not function as a "catalyst" or as a "rapid data synthesis method" to bring about more rapid resolution of controversy. Indeed, since the growth of evidence-based methods in the early 1990s, such a function would have been viewed with suspicion because systematic evidence reviews are the currently preferred way of synthesizing complex data.

The NIH Consensus Development Conference Program, even in the time before evidence-based medicine methods became a part of the process, did not function to develop a "technical consensus" because it usually missed the intended window of epistemic opportunity. NIH Consensus Conferences took over a year to plan and intentionally picked topics where they thought there was sufficient evidence to come to an

---

[35] E-mail communication from David Murray via Elizabeth Neilson, May 21, 2014.
[36] Recall that "technical" consensus is consensus on scientific matters; "interface" consensus is consensus on ethical, political, economic, and social matters connected with the use of the technology.

evidence-based consensus (rather than simply agreement for the sake of unanimity). So they typically took place *after* the experts reached consensus, i.e. *too late to produce a technical consensus*. For example, the 1994 NIH Consensus Development Conference, Helicobacter Pylori in Peptic Ulcer Disease, took place after the important clinical trials (in the late 1980s and early 1990s, some sponsored by the NIH) and after research scientists and many prominent clinicians reached consensus on the use of antibiotics for peptic ulcers. The 2002 NIH Consensus Development Conference, Management of Hepatitis C: 2002, repeats recommendations that were stated by the Food and Drug Administration (FDA) the previous year.

John Ferguson, the longest serving director of OMAR (1988–99), wrote in 1993: "Often the planners of any given consensus conference are aware of a likely outcome and use the conference as a mechanism to inform the health care community" (Ferguson 1993: 189). This is an official acknowledgment that consensus conferences typically took place too late to resolve controversy among researchers. The "dissemination of knowledge" goal remained the justification for going through the process. But this means that the specific recommendations of an NIH Consensus Development Conference were not really determined by the process of the consensus conference; instead, they were predetermined by the state of research prior to the conference. Since 2001, the state of research prior to the conference has been summarized in an AHRQ evidence report, and recommendations of NIH Consensus Development Conferences were governed by its results.

If we really think about this, it is not a disappointment. We would not want consensus conferences to produce consensus when there is insufficient evidence that consensus is justified. A forced early consensus would not be in the interests of truth, patient health, or the credibility of the NIH Consensus Development Conference Program. Early on, it was hoped that the NIH Consensus Development Conference Program would *accelerate* the usual process of forming consensus in science (Asch & Lowe 1984): i.e. keep the consensus evidence-based but make it happen faster. It turned out, however, that the time frame of an NIH Consensus Development Conference (from planning to final statement, at least a year) was generally longer than the time frame of the usual scientific progress, and so was not able to accelerate it. Moreover, the NIH Consensus Development Conference Program, in later years, did not want to bypass or shortchange the process of systematic evidence assessment. If a controversy cannot be resolved by the evidence report it should not be resolved by other means, on pain of making the process non-evidence based.[37]

The "dissemination of knowledge" goal remained an official goal of the program (although it seems to have been forgotten at the very end), and was taken seriously, as shown by repeated studies of the impact of NIH Consensus Development Conferences and attempts to strengthen the impact through wider distribution of results, webcast streaming of conferences, and a comprehensive website. More precisely, the goal was *practice change* as a result of knowledge dissemination. Practice change is notoriously slow—on the order of several years for many medical innovations—and therefore compatible with the time frame of an NIH Consensus Development Conference. A well-timed NIH Consensus Development Conference could accelerate the usual process of dissemination leading to practice change, and there are a few cases in which a little of this seems to have happened (Ferguson 1993). Usually, however, it took more than an NIH Consensus Development Conference to effect practice change.[38] (Consensus conference programs in other countries have been more effective at influencing practice change, as will be discussed in Chapter 3.)

Knowledge dissemination and practice change are social epistemic goals which take place not by mere transmission of the results of the original researchers, but by additional means such as trust of the original researchers, peer pressure, pressure from trusted authorities, and incentives for change. I see the NIH Consensus Development Conference Program as an attempt to construct a *trusted source of information*—especially trusted by clinicians and the general public—about a variety of new health care interventions. John Ferguson's metaphor of the NIH Consensus Development Conference Program as "NIH's window to the health care community" says it well, if we elaborate that the window is selective and shows only the results deemed trustable.

---

[37] Some consensus programs allow expert consensus to fill in the blanks that evidence does not reach (see Chapter 3). This could in theory be a role for the NIH Consensus Development Conference Program, but because panelists are not experts in the area (they are usually experts in a related area) they do not comprise the "expert consensus" needed. In any case, expert consensus is typically on the lowest rung of the evidence hierarchy.

[38] When there are financial incentives for practice change, things go more quickly.

Like all knowledge, scientific knowledge is communicated on trust, and trust is all about the *perception* of a source as trustable.[39] If a source is in fact trustable, but does not *look* trustable, then the knowledge will not be communicated. Moreover, if a source is not trustable, but seems trustable, its knowledge can be disseminated. So the *appearance* of trustability is what matters for the communication of scientific knowledge. A source appears trustable when it is judged as honest, unbiased, knowledgeable, etc. Of course, different audiences may make different judgments about trustability, so sources may need to prove themselves in different ways to those different audiences. The word "objective" is often used to describe the source of trustable knowledge.

"Objective" can mean many different things. In the case of the NIH Consensus Development Conference Program, it is an assessment of the process used in making a consensus statement: an independent panel, the names and physical presence of qualified panelists, emphasis on openness of proceedings, presentations from experts on all sides of the issue, careful discussion, *gravitas* of the proceedings, and the imprimatur of the NIH, which is the largest and most prestigious health care research facility in the world. The secrecy of executive sessions keeps any negotiations which may look "unscientific"[40] hidden. The process is designed to look objective to its intended audience—health care providers and the public. "Objective" is understood both negatively (as "freedom from bias") and positively (as "a group of experts following careful deliberation").

The NIH Consensus Development Conference Program worked hard to respond to any allegations of bias by attempting to improve the perceived objectivity of the process. Sometimes these efforts required a juggling act, as shown in the epistemic tensions described in Section 2.5. Sometimes not all the intended audiences could be satisfied: for example, audiences that were sophisticated about group process were not satisfied by unstructured discussion, but audiences that were unsophisticated about group process welcomed unstructured discussion because it seems unconstrained and therefore "free." Objectivity is an ideal that is never perfectly achieved, even for a homogeneous audience and much less for a non-homogeneous audience.

[39] "Trustable" and "reliable" do not mean "indubitable." There is room for challenge, so long as not everything is challenged.
[40] It is a problem for credibility when negotiations look unscientific, whether or not they are indeed unscientific.

Fortunately, credibility and trustability do not demand perfection. (On the other hand, distrust makes the most of any imperfection.) Moreover, variations in the process of coming to consensus will not have important consequences because the outcome of the process (the consensus statement) is not much influenced by the process of deliberation. As already discussed, the outcome of the process is generally predetermined by the consensus of the research community and/or by the AHRQ-sponsored evidence report.

The appearance of objectivity can be maintained, despite some minor concerns about bias, so long as consensus statements remain authoritative, at least for a while (as long as the statements are consistent with the latest evidence). There is no great risk with granting them this authority, since the research community has already reached the same conclusion on the matter and/or the evidence report has reached the same conclusions. The reliability of NIH Consensus Development Conferences was the result of the correspondence of the consensus reached at the conference to the consensus reached in the research community and/or the results of the AHRQ-sponsored evidence report. For this reason, it would have been pointless to test the reliability of NIH Consensus Development Conferences by running more than one panel simultaneously and seeing whether or not similar results are achieved, or by comparing the results of the panels with other methods such as formal syntheses.[41]

Different consensus development programs may work with a different balance of epistemic tensions. For example, as I will show in Chapter 3, IOM consensus panels, unlike the NIH Consensus Development Conference Program, do not trade expertise for neutrality; they prefer to "balance biases" rather than eliminate them and lose expertise as a result, so they invite the best known and published experts in a field to serve on a panel and hope that they avoid bias by issuing a good balance of invitations. The IOM also prefers a reflective decision, which may take some time while increasing the credibility of the result, over the rhetorical impact of a consensus statement given in a press conference on the third day of the conference.[42] Other consensus programs make other choices.

---

[41] It is worth testing the reliability of other consensus conference programs for which conclusions are not so predetermined. Examples include Scott and Black (1991) and Leape et al. (1992), which found that different panels can come to different conclusions on the same questions.

[42] John R. Clarke tells me that the IOM also does a good job of creating rhetorical impact with a press conference on release of a report. It may be that the drama of a three-day event with a press conference at the end is not necessary for impact.

Even for the NIH Consensus Development Conference Program, other choices could have been made along the way (e.g. for more structure in group deliberation, or more oversight of the process by an external committee). Because different balances of epistemic tensions can also work well, we can conclude that the NIH Consensus Development Conference Program had a good deal of historical contingency.

The most plausible justification for continuing with NIH Consensus Development Conferences—as well as the reason for beginning them in the first place—was because they produced credibility and they translated research findings into a form in which a general health care provider or member of the public could understand. Systematic evidence reviews are produced by small groups of statisticians, information retrieval specialists, and research methodologists; in the form that they are produced, they are not immediately grasped as credible by the general public or the general health care practitioner.

I suggest that NIH Consensus Development Conferences were *social epistemic rituals* which conferred a stamp of epistemic approval on results of research and translated the results in a way that was easily comprehensible to a broad audience. Their traditional methods and careful choreography contributed to their perceived credibility and objectivity. Despite being called "consensus development conferences," they did not really process information independently of the research community. They are not like the magazine *Consumer Reports* for example, which makes its own determinations and collects its own data. Moreover, for effectiveness their perceived objectivity matters more than their actual objectivity.

It remains to ask, how successful were NIH Consensus Development Conferences at disseminating knowledge and changing practice, and what is the loss in their "retirement"? One study shows that the most important factor, after prior practice, influencing practitioners' responses to medical consensus statements is the "perceived influence" of the members of a consensus panel itself (Hill 1991), for which the NIH Program did well, selecting panelists and chairs with strong professional reputations. This result is consistent with my assessment that it is the credibility—the rhetorical force—of NIH Consensus Development Conferences that matters. Nevertheless, most studies of the impact of NIH Consensus Development Conferences show little impact on clinical practice (Ferguson 1993, Jacoby 1983, Kosecoff et al. 1987, Kosecoff et al. 1990, Jacoby & Clark 1986, Thamer et al. 1998, Lazovich et al. 1999).

On the occasions when they seemed to have impact, other causes of change of practice (such as financial incentives) may have played a facilitating role.[43] This disappointing result needs to be put in the context of decades of continuing frustration with unsuccessful attempts to change clinician practice through knowledge dissemination, more accurately called "passive knowledge dissemination." Financial or social incentives may be more effective, or may increase the effect of knowledge dissemination in clinical practice. Atul Gawande has recently argued that clinical practice is best disseminated through one-on-one social interaction (Gawande 2013). The NIH Consensus Development Conference Program was not designed to do this.

Although the NIH has emphasized two purposes of its consensus conferences—to bring about consensus and to disseminate it—the NIH Consensus Development Conference Program may have had other functions which are not mentioned (maybe not even noticed), or not mentioned as often. One, which was recently noted (Portnoy et al. 2007), is the role of NIH Consensus Development Conferences in influencing further research. From the beginning, the NIH Consensus Development Conference Program aimed to identify those areas of research that are inconclusive and need further attention. The original purpose was to warn clinicians and the public that a particular health care intervention is not ready for prime time implementation. Portnoy et al. (2007) shows that NIH Consensus Development Conferences also have a positive effect, in that they have influenced the number of research proposals submitted to the NIH on particular topics from 1998 to 2001.

Another important function of NIH Consensus Development Conferences may be their ability to prevent or dissuade other interested groups from putting out competing claims. It is more difficult to sustain or manufacture doubt (which interested groups such as health insurance providers can do) when there is an authoritative source of knowledge. The rhetoric of expert consensus is powerful: if 20 eminent gastroenterologists agree on a gastroenterological matter, you disagree at your own peril.

---

[43] For example, the 1983 conference on liver transplantation had a large impact on practice, but the impact on insurers (who agreed to cover the procedure) may have been the critical step. See Ferguson (1993).

In "retiring" the NIH Consensus Development Conference Program, the NIH hopes and expects that other groups will take up any slack. To some degree, they are right in thinking that they provided the model for other groups to make assessments of new medical technologies. Perhaps the use of the word "retire" is intended to imply the passage of the same task to younger organizations that NIH inspired. But this misses the important influence of the evidence-based medicine movement on the process of assessing new medical technologies. The task has changed—from group knowledge assessment and dissemination to formal knowledge assessment techniques and other efforts (instead of or in addition to consensus conferences) for the difficult process of dissemination.

## 2.8  Conclusions

NIH Consensus Development Conferences were choreographed social epistemic rituals, designed to produce conclusions that were perceived as "objective" by its intended audience. This intended audience—clinicians and the general public—is unsophisticated about both group process and, more recently, about evidence synthesis. The NIH Consensus Development Conference Program made use of the rhetorical appeal of a consensus of experts to produce knowledge that was authoritative for its audience. The perceived conduct of the process was more important for the trustability of the outcome than the content of the resulting statement, which was usually predictable from an already existing consensus of experts. The precise form of the conferences had some contingency. Changes in format over the years were due to attempts to balance both old and new epistemic tensions over the handling of bias, timing of the conference, time pressures, specificity versus generality, and technical versus interface issues. The greatest challenge to the NIH Consensus Development Conference Program was the development of evidence-based medicine, and eventually the program responded to this challenge, surviving for another decade.

NIH Consensus Development Conferences should be evaluated not in terms of their ability to move the scientific community from disagreement to consensus (which, clearly, they did not accomplish), but rather in terms of their broader epistemic achievements, especially the tasks of knowledge dissemination, practice change, maintenance of epistemic authority, and stimulation of particular kinds of research. Their impact, so far as we have been able to tell, has been modest, but we could not have known this in advance. We are working on (but do not yet have) more powerful tools to accomplish these tasks.

Will the NIH Consensus Development Program be missed? The fact that its "retirement" has gone unremarked in print (except here) suggests that it no longer has champions. But I do *not* draw the conclusion that consensus conferences have been superseded by evidence-based medicine. We have not lost the need for the epistemic authority of group consensus. It is in this respect—unfortunately not acknowledged by those who ended it—that the program can be "retired" having left a continuing legacy of other medical consensus conferences.

# 3

# From the NIH Model to the Danish Model

The Medical Consensus Conference Movement

## 3.1 Introduction

After its creation in 1977, the NIH Consensus Development Conference Program immediately became a model for the development of other medical consensus conference programs, both in the United States and in other countries.[1] This influence preceded any positive formal evaluations. Process evaluations took five or so years to produce and outcome evaluations (formal assessments of reliability and validity) of the NIH Program have never been done.[2] Effects on clinical practice took some time to evaluate (Jacoby 1983, Ferguson 1993). Rather, the *process* of group deliberation *made sense* to all but a few detractors, and the *experience* of participation in group deliberation was mostly positive for panelists.[3] Finally, the outcomes were usually acceptable, agreeing with the best current research.[4] Other consensus conference programs designed

---

[1] My essay "Group judgment and the medical consensus conference" (Solomon 2011a) was a preliminary version of some of the material in this chapter.

[2] If we think about what it would take to do outcome evaluations, it is not surprising that they have not been done. We would need a firm measure of "truth," or at least of what the best decision is. But that is precisely what is in question when a consensus conference is held.

[3] These points were covered in Chapter 2. I will say more about group deliberation "making sense" in Chapter 4.

[4] As I argued in Chapter 2, this result may have been predetermined by the fact that consensus conferences take about a year to schedule, and by the time they meet there may no longer be any controversy among researchers.

on the NIH model were similarly well received. By 1990, an IOM report observed that "Group judgment methods are perhaps the most widely used means of assessment of medical technologies in many countries" (1990: 1).[5]

The goal of this chapter is to describe and analyze the social epistemology of these other (non-NIH) medical consensus conferences in the United States and internationally. The different consensus conferences developed different stable forms with different balances of epistemic tensions,[6] contributing to actual and perceived objectivity and fairness. It turns out, in fact, that although the NIH Consensus Development Conference Program is the original model for consensus conferences, it is not typical of medical consensus conferences, even in the United States. Moreover, another model for consensus conferences developed, owing much to the NIH Program, but with a different resolution of epistemic tensions. This is the "Danish model," which is a model of public democratic deliberation rather than a model of group expert judgment.

In Chapter 2, I found that the NIH Consensus Development Conferences do *not*—their name notwithstanding—process information to develop a consensus about a medical technology when researchers dissent. I argued that, instead, they function (at best) to make a prior consensus among researchers authoritative and trustable, and to disseminate it to the wider community. Other medical consensus conference programs likewise do not assess or process information to develop a consensus when researchers dissent. These other medical consensus conferences may have the same function as the NIH and/or they may be an opportunity for negotiation of policy and "buy-in" of targeted communities and individuals, as I will show in the course of this chapter.

The greatest difference between the NIH and all other consensus conferences (both in the United States and internationally) is that NIH Consensus Conferences represent an extreme in their focus on "technical" (meaning scientific) consensus rather than "interface" (meaning broader implications) consensus.[7] NIH's emphasis on considering

---

[5] Note that the epistemic situation has changed since that time. Now the IOM would say that the techniques of evidence-based medicine are the most widely used means of assessment of medical technologies.

[6] These epistemic tensions were described in Chapter 2.

[7] Recall from Chapter 2 that from its beginnings the NIH Program distinguished "technical" from "interface" consensus, and stated its interest in achieving only the former kind of consensus.

technical questions permitted me to focus in Chapter 2 on the epistemic aspects of consensus conferences. Other United States and international medical consensus conferences regularly consider broader ethical, political, economic, and social implications of medical research and practice. They explicitly and unapologetically engage in what the NIH calls "interface" consensus in addition to "technical" consensus.

The greatest difference between North American and European consensus conferences is that North American conferences have panels of experts (with occasional lay representatives), while European consensus conferences often have entirely lay panels (with experts advising them).

So far as I am aware, the medical consensus conference movement was and is a First World phenomenon. My focus on North America and Northern Europe (within the larger First World category) is a reflection of the prominence of medical consensus conferences in these countries. Additionally, my resources for information about non-US medical consensus conferences have been thin, mostly consisting of the published literature in English. This is a linguistic and geographic bias that I hope is remedied by other researchers.

I will not be able to describe and discuss other consensus conferences in as much detail as I did for the NIH Consensus Development Conference Program. To do an adequate job of tracing the development of these conferences over time requires archival research that is beyond the scope of this book. Such detail is also not necessary for making the comparisons that I focus on. I will present a summary account that highlights the significant differences between consensus conference programs and shows their developments over time.

## 3.2  Medical Consensus Conferences in the United States

NIH Consensus Conferences were the model for other medical consensus conferences. Early criticisms of their objectivity were taken to heart by those devising other consensus conference programs, especially in the US, presumably because those in the United States had direct experience of both the benefits and the drawbacks of the NIH model. These other US medical consensus conference programs did not replicate the NIH model, but modified it to better correspond with the conception(s) of objectivity most pertinent or practical to them. In particular, they deal differently with panel composition, time pressure, oversight, technical versus interface consensus, degree of specificity of recommendations, and definition of consensus.[8] That is, they balance the epistemic tensions of consensus conferences differently.

In Chapter 2, the epistemic tensions for the NIH Consensus Development Conference Program were listed as: (a) there should be a balance between relevant expertise of panelists and the need for perceived impartiality of conclusions; (b) the timing of consensus conferences should be not so early that there is an insufficient evidence base, but not so late that there is already consensus in the wider community; (c) the length of consensus conferences should not be too short to consider matters properly, yet not be so long that panelists will be reluctant to participate or costs will be too high; (d) consensus statements should be specific enough to avoid being vacuous, but not so specific that they look like they are prescribing the practice of medicine; (e) there are choices about focus on technical versus interface concerns; (f) some external review increases confidence in the conclusions, but too much undermines the idea that the conference does the important work; and (g) the conferences should be open enough for trust in their conclusions, but private enough to get the work done and to maintain an air of authority.

One of the most respected medical consensus programs in the United States is run by the IOM, and is a good example of modification of the NIH model to produce an arguably more attainable and useful kind of objectivity.[9] IOM Consensus Panels do not aim for the elimination of "intellectual bias," instead aiming for a "balance" of expertise on panels. The IOM distinguishes "point of view" from "conflict of interest," implicitly suggesting that the latter only is epistemically objectionable.[10] They judge the loss of domain expertise and the absence of interested parties to be more detrimental than the gain in neutrality obtained by excluding

---

[8] Definition of consensus may be complete unanimity or majority (with or without a minority statement).

[9] Information about the IOM Consensus Study program is taken from the IOM website <http://www.iom.edu> and from the National Academies website <http://www.nation­alacademies.org/studyprocess/index.html>, accessed July 20, 2014.

[10] Of course, conflict of interest is morally objectionable, while having a point of view is morally neutral and an unavoidable result of high quality intellectual engagement. Confusion of the epistemic with the ethical is common in discussions of conflict of interest.

scientists with published work in the area of the conference. The NIH Consensus Development Conference Program implicitly made the contrary decision by excluding those with what they call "intellectual bias." No one has formally tested the comparative reliability of these methods for technical consensus, or for producing trust in the results and dissemination of conclusions.[11] It is likely that the IOM model is superior for interface consensus, since it brings the involved parties (rather than neutral bystanders) to the table. The NIH model is particularly unsuited to interface consensus, since it brings neutral bystanders rather than the involved parties to the table.

The IOM, like the NIH, has information-gathering meetings open to the public and deliberative meetings in private. (I discussed the tension between openness and privacy in Chapter 2.) Unlike the NIH, the IOM does not attempt to come to conclusions in two-and-a-half days: it takes as many meetings as are necessary, typically around three, plus conference calls and subcommittees, spread over a period of around a year. Perhaps the prestige of the IOM, as well as the stake that each panelist has in the conclusions (because they are active researchers rather than neutral bystanders) helps to persuade panelists to devote this substantial time. Finally, the result of deliberation (which can include continuing dissent) is reviewed by a second, independent panel of experts whose comments are provided anonymously to the committee members. The IOM process was also shaped by the membership of IOM in the National Academy of Sciences, which emphasizes the important role of expert selection and peer review in all its projects. The IOM is typically charged by Congress or federal agencies,[12] with the implication that legislation will follow.[13] This contrasts with the NIH Program, which preferred not to influence legal matters.

Another example of similarity and difference with the NIH Consensus Development Conference Program is the work of the Medicare Coverage Advisory Committee (established in 1998, renamed in 2007 as the Medicare Evidence Development Coverage Advisory Committee or

---

[11] I suspect that dissemination will be better when the experts in the area sign on to the consensus, either as panelists or as reviewers.

[12] See <http://www.iom.edu/About-IOM.aspx>, accessed July 29, 2014.

[13] I am grateful to John R. Clarke for pointing this out to me.

MedCAC). The goal of the committee is to "allow an unbiased and current deliberation of 'state of the art' technology and science"[14] so that they can advise the Center for Medicare and Medicaid Services (CMS) on the newest therapeutic interventions. These days, their deliberations are preceded by a formal evidence assessment (typically performed by one of the evidence-based practice centers under AHRQ). Their charge is to determine how to apply the evidence to the Medicare population. Panelists are non-governmental employees, and mostly physician and researcher experts in the medical fields under discussion. Each panel of 13–15 people also includes a patient advocate, consumer representative, and industry representative. After discussion, panelists vote individually on each of a large number of questions and do not require the panel to debate to unanimity (which is usually what consensus means, and certainly what consensus means in the NIH Program). Votes become part of the public record. MedCAC works under the same kind of time pressure as NIH, devoting one meeting to each topic. Note that MedCAC still values face-to-face meeting and group discussion, so it still values group deliberation and even uses the term "deliberation" in its mission statement.[15] So MedCAC does not simply aggregate the initial opinions of a group of individuals, nor does it strive to produce unanimity; its results are somewhere in between an aggregation of initial individual opinions and a discussion producing consensus. It reflects a more sophisticated understanding of the epistemic pitfalls of group process than the NIH and many other consensus conferences have. (Chapter 4 will go into more detail about group process and the social epistemology of group deliberation.) My impression is that those consensus conferences intended for more sophisticated audiences have more sophisticated group processes; the MedCAC conferences are not intended for the public or for primary care practitioners, but report to a government body, the CMS. It should also be noted that taking a vote is easier and quicker than attempting to come to consensus; this may suit the pragmatic goals of MedCAC. It runs the risk, however, of making the results less persuasive: if the experts are not in agreement, there may be reason for doubt. Taking a vote also

---

[14] From <http://www.cms.gov/Regulations-and-Guidance/Guidance/FACA/MEDCAC.html>, accessed July 20, 2014.

[15] The mission statement is at <http://www.cms.gov/Regulations-and-Guidance/Guidance/FACA/MEDCAC.html>, accessed July 20, 2014.

brings an end to deliberations; there are no further stages of review (as there are with the IOM). Time pressure may negatively affect decisions, but the pressure is not as great as with the NIH Consensus Development Conference Program, where a consensus document needs to be produced in three days; MedCAC only has to produce the vote count.

MedCAC, in the same way as the IOM, includes experts on the panel with what the NIH Consensus Development Conference Program would call "intellectual bias" rather than independent semi-experts. Bringing the experts into the discussion (despite their "intellectual biases") plausibly results in more buy-in by those experts participating and by those who respect the participation of particular experts (e.g. clinicians who see a major researcher in the field represented on the committee).

The US Preventive Services Task Force (USPSTF) began in 1984 as a consensus conference program. Like other consensus conference programs, its evidence base has improved over the years. Since 1998, it has been supported by AHRQ and two evidence-based practice centers, which produce systematic evidence reviews before the panel meets. It produces clinical practice guidelines for preventive care, and currently describes itself as "an independent panel of non-Federal experts in prevention and evidence-based medicine."[16] The USPSTF has 16 volunteer members, who are physicians or statisticians with expertise in primary care, preventive medicine, and evidence-based medicine. They form a long-term committee, basically the same committee for each consensus conference. They meet several times per year and communicate between meetings, working on a number of guidelines at the same time. The USPSTF partners with many professional, policy, and governmental organizations (from the American Academy of Family Physicians to the American Association of Retired Persons (AARP) to the Canadian Task Force on Preventive Health Care[17]) to review the recommendations. The conflict of interest policy has been strengthened recently to require that members of the USPSTF give "full disclosure of their interests."[18] The USPSTF proceeds by estimating the magnitude of harms and benefits of

[16] <http://www.ahrq.gov/professionals/clinicians-providers/guidelines-recommendations/uspstf/index.html>, accessed July 20, 2014.
[17] The full list of partners is at <http://www.uspreventiveservicestaskforce.org/Page/Name/our-partners>, accessed July 20, 2014.
[18] <http://www.uspreventiveservicestaskforce.org/Page/Name/procedure-manual-section-1>, accessed October 30, 2014.

the particular preventive service and then voting on clinical recommendations. If consensus is not reached, there are no minority reports (perhaps this is to avoid any loss of authority). Economic and other societal factors are taken into account. Drafts of statements are posted online for public comment for four weeks. Statements also go through review by partner organizations, which in turn help with dissemination.

The USPSTF is designed to achieve both technical and interface consensus. Its partnering with organizations such as AARP shows both that the USPSTF takes interface questions seriously and that it works at getting "buy-in" from influential organizations. Furthermore, it asks for review from many different organizations and experts, and review is both a check and a way of getting "buy-in" for recommendations. The inclusion of review means that that there is no public announcement of the consensus after the panel meets. This program prefers to take adequate time for reflection and revision than to get attention with a press conference held at the conclusion of discussions. The use of voting makes it easier to come to conclusions, which probably enables the USPSTF to take on more projects. The cost, of course, is that the USPSTF cannot always claim to have complete unanimity, which reduces the rhetorical power of its recommendations.

These three examples—the IOM, MedCAC, and the USPSTF are typical examples of medical consensus programs in the US. Other well-known and respected general programs are the Blue Cross and Blue Shield Technology Evaluation Center Assessments and the American College of Physicians' Clinical Efficacy Assessment project. Medical specialty groups such as the American College of Radiology, the Society of American Gastrointestinal and Endoscopic Surgeons, and the American Society of Clinical Oncology also produce guidelines and recommendations that claim to be both evidence based and consensus based.

Sometimes different organizations produce guidelines that conflict with one another. A well-known example is the series of controversies during the past 20 years over the use of breast cancer screening mammography for women aged 40–49, in which the American Cancer Society and the American College of Radiology have recommended routine screening and the NIH Consensus Development Conference Program and the American College of Physicians have not recommended it. Such lack of consensus between consensus conferences weakens the perceived objectivity of at least some of the sources of the recommendations. When objectivity is questioned, specific accusations of bias are not far behind. The American College of Radiology, for example, has been accused of self-interest in its recommendation of an annual screening mammogram for all women over 40 (Quanstrum & Hayward 2010). This case will be discussed in detail in Chapter 9.

Since the late 1990s, most consensus programs have revised their procedures, and sometimes their names, in response to the growing importance of evidence-based medicine. Several (e.g. the NIH Consensus Development Conference Program until it ended in 2013, the American College of Physicians' Clinical Efficacy Assessment project, MedCAC, and the USPSTF) work in conjunction with an AHRQ Evidence-Based Practice Center (the main producer of systematic evidence reviews in the United States), commissioning an evidence report prior to the conference or meeting. So it is now generally accepted that consensus conferences are not for group assessments of complex evidence; this work of evidence assessments is done more formally by an earlier evidence review.

Why have consensus conferences continued when their original rationale—doing a complex evidence assessment—clearly no longer applies? This question arose for the NIH conferences after 2001, and in Chapter 2 I mentioned the claim of NIH personnel during the 2000s that the current function of the consensus panel is to "extrapolate" in justified ways from the available evidence. Alan Sniderman (1999) argues more specifically that consensus conferences are needed to make decisions about the external validity of clinical trials.

These rationales are no more convincing for non-NIH Consensus Conferences than they were for NIH Consensus Conferences. The panel format—face to face, 10–20 people who are seen as independent, trustworthy, and objective, deliberating and coming to agreement—is not especially well suited to tasks of "extrapolation" or determination of external validity, which can be done more easily by the domain experts involved with producing the evidence synthesis (with external review if necessary). Translation into non-technical language (if required for the intended audience of the consensus report) can be done by professional writers. I argued in the case of NIH that the main purpose of the conferences is to lend authority to the results, giving the results faces, signatures, and handshakes. This is rhetorically powerful even and perhaps especially in the age of evidence-based medicine, when there is (deservedly and/or undeservedly) both lay and professional skepticism about systematic evidence review and meta-analysis. The non-NIH confer-ences also produce authoritative statements, amplified by buy-in from those experts who participate either in the conference or in the review process. Finally, and perhaps most important, a deliberative consensus on ethical and policy matters, i.e. interface consensus, requires the group interaction of a consensus conference.

## 3.3  Canadian and European Consensus Conferences

Health care provision in the United States is decentralized. Although the government provides health care to some poor people and to the elderly, most people's care is provided by private health care insurance, and health care is not (yet) universal. In Canada and Europe, on the other hand, health care is universal and centralized. This has important con-sequences for the impact of consensus conferences in these countries. Hence, I put Canada and Europe in the same section for this part of the discussion.

The Canadian Task Force on Preventive Health Care (Canadian Task Force, established 1979, discontinued in 2005, and restarted in 2010) pro-duces guidelines for use in primary care medicine.[19] Like the USPSTF (which it influenced), it has a permanent panel, rather than different panels for different topics. Obviously this is more efficient than recruit-ing and training a new panel for each topic, and it makes sense because the emphasis of the Task Force is on primary care and the 14 panelists are therefore generalists. In contrast, the NIH Consensus Development Conference Program had a new panel for each consensus conference. This was necessary because the NIH often evaluated specialty care, and its panelists needed to be specialists on the topic under discussion (or as specialist as possible while avoiding "intellectual bias"). A consideration raised by Sniderman (1999) about consensus conferences generally is that if the panel is permanent, there may be "inordinate linkage between indi-viduals and the process." His concern is that personal dynamics will affect the outcome of the group process. This leads Sniderman to recommend

---

[19] The date of origin suggests that the Canadian Task Force may have been created independently of the NIH Consensus Development Conference. Despite a number of attempts, I have not yet discovered much about its origins.

that a "substantial portion" of new participants should be used for each conference. Interestingly, in both the USPSTF and the Canadian Task Force members serve for a maximum of four years in staggered terms, achieving a kind of compromise between the temporary and the permanent. Here is another area in which epistemic balance is important.

A 1990 article on the Canadian Task Force reported that individual panel members were responsible for review of the literature, and that arriving at consensus "requires from four to eight iterations spread over one to two years" of meetings (Battista 1990: 89). After this, external peer review evaluated the recommendations of the Canadian Task Force. This was a highly regarded program for a long time; the 1994 volume *The Canadian Guide to Preventive Health Care* (the "red brick") became the standard reference for Canadian primary care physicians. The version of the "red brick" in France was closely influenced by the Canadian version, and the USPSTF modeled itself on the Canadian program. (Because health care is less centralized in the US, it is difficult for anything to become the "standard reference.") The emphasis on consensus in earlier reports was supplemented during the 1990s with an emphasis on evidence-based decision making. Canada is one of the birthplaces of evidence-based medicine (the other place being the UK), so it is not surprising to see a move to add a formal evidence review. The Canadian program still exists; it was restarted in April 2010 with support from the Public Health Agency of Canada, after stopping in 2005 (the reasons for this are unknown to me). It now has an associated Evidence Review and Synthesis Center at McMaster University (the birthplace of Canadian evidence-based medicine) and a "Task Force Partners Group," which is responsible for dissemination and implementation of results. Because of the centralization of health care in Canada and the emphasis on dissemination and implementation, the guidelines of the Canadian Task Force have had considerably more impact than the USPSTF.

The first European consensus conference took place in Sweden in 1982 on the topic of hip replacement surgery. It was set up in consultation with NIH personnel and in the wake of early enthusiasm about consensus conferences. As already mentioned in Chapter 2, satisfaction was expressed at its conclusions because the final statement agreed with the conclusions of an NIH Consensus Development Conference held two months previously on the same topic (Rogers et al. 1982). Reaching the same conclusions is hardly surprising, especially given the connection between the two programs. Furthermore, why did they bother to repeat the hip replacement consensus conference in a different country? Since the technology is the same, and the population health similar (both industrialized nations), why not save time and effort and use the results of the NIH conference for Sweden? The justification given in 1982 was that the Swedish conference focused on need and cost (interface consensus questions), whereas the NIH Consensus Development Conference eschewed such considerations and focused only on technical (scientific) matters. But still, why repeat the scientific assessment (the technical consensus)? Why not just focus on the discussion of economics, ethics, and policy (the "interface" consensus)? Per Buch Andreasen proposes that the credibility of each consensus has to be established for each intended audience:

> From a democratic point of view, the process itself is the most important element in achieving credibility and legitimacy for the results. This is another reason why technology assessments performed in one country will have limited impact for other countries. (Andreasen 1988: 306)

This suggests that it is not enough for the Swedish context that the NIH Consensus Development Conference reached consensus on the effectiveness of a particular technology—even when there is no specific reason for skepticism about the NIH Consensus Development Conference process. Epistemic trust, even of the NIH, is not so strong that it crosses borders of language and culture without the need for amplification.

In the 1980s, several countries developed their own consensus development programs for medical technology. Some, such as Denmark, were so impressed with the method that they used it more broadly for a number of questions in biotechnology, eventually transforming the process into the "Danish consensus conference" model and then exporting it back to the US and elsewhere (Guston 1999).

Most of the US and Canadian consensus conference programs (other than the NIH, of course) provide ample time for reflection and make no public announcements at the end of the conference. Not so in Europe. Perhaps the Europeans' relative lack of experience with the NIH Program made them less aware of the deficits of time pressure, and/or perhaps time pressure is more helpful, or rhetorically appropriate, in a more public "interface" consensus. I will say more about this later in the section.

In Europe, and especially in Scandinavian countries, the focus of medical consensus conferences tends to be more on "interface" than on "technical" consensus, in that they encourage discussion of economic, ethical, and political factors. Furthermore, their aim is to inform the public and politicians, rather than clinicians. And, most distinctive, and in contrast to the NIH model of expert group judgment, they have been adopted as a model of public democratic deliberation. Correspondingly, the panels have much higher representation from the public and from non-medical professionals. In addition, European consensus conferences do not undergo another level of expert review. This is because of the emphasis on democratic deliberation about practical matters. Experts who criticize and try to shape the final statement would vitiate the process.

For example, the Danish Medical Research Council, which began consensus conferences in 1983, included journalists, politicians, and patient representatives on the panel (Jorgensen 1990). The Norwegian National Research Council began a consensus development program in 1989. It only took on topics with significant ethical or social consequences (Backer 1990). The Swedish consensus development program (established late 1981) took on questions about cost implications from the beginning (Jacoby 1990). All the European medical consensus conference programs were modeled on the NIH Consensus Development Conference Program and preserve some of its controversial features such as requiring a neutral (rather than a "balance of biases") panel, one and a half days of expert presentations with questions, and a public statement on the third day (with overnight work by the panel if necessary).

As I suggested above, perhaps the European programs adopted the NIH process with fewer changes than consensus programs in the United States because Europeans did not gain early experience from direct participation in the NIH Consensus Development Conference Program. Many of the same (usually North American) individuals serve on a wide range of consensus panels in the US.[20] Perhaps those who have to stay up all night writing a consensus statement are more aware of the possibility of a biased outcome than those who have not had that experience!

---

[20] To mention just two individuals that I have spoken with about their participation: Sanford (Sandy) Schwartz (University of Pennsylvania) and Hal Sox (Editor Emeritus, *Annals of Internal Medicine*) have participated at meetings of almost all the United States consensus programs discussed in this chapter.

Consensus conferences on non-medical issues (e.g. environmental action and international aid priorities) developed in Europe as an expansion of the institution of medical consensus conferences, especially in Scandinavian countries, where they are referred to as "the Danish model." The method of consensus conferences with public participation coheres with the style of participatory democracy that is politically established in these countries (Horst & Irwin 2010).

European and Canadian consensus conferences have, in general, been more successful than United States consensus conferences with dissemination of their conclusions (Calltorp 1988, Johnsson 1988). The European and Canadian consensus conferences tend to be organized by governmental bodies, or organizations that report directly to them, and the results of the consensus conferences are directly fed into a centralized bureaucracy for health care provision and reimbursement. Not so in the United States, where health care is more decentralized and physicians have more autonomy. Dissemination of conclusions is not the same as practice change, since physicians can be aware of conclusions without adopting them in their own practices. But dissemination it is a step toward practice change. Moreover, linking health care regulation and reimbursement to the outcomes of consensus conferences has provided additional incentives for practice change.

In summary, there are many consensus conference programs, some public (such as the NIH Consensus Development Conference Program), some professional (such as the Eastern Association for the Surgery of Trauma consensus conference), some private (such as the Blue Cross and Blue Shield Technology Evaluation Centers), some Northern European (such as the Danish Medical Research Council) and some international (such as the International Consensus Conferences in Intensive Care Medicine).[21] They have different emphases on technical versus interface consensus and they differ in some details of their procedures, but all have in common a reliance on the presumed rationality and fairness of properly conducted "group process." The epistemic appeal is in the idea of an expert and/or representative panel coming together, face to face, to sort out the issues. When the issues are economic, political, or ethical as well as scientific, experts from these applied fields (e.g. ethicists,

---

[21] International consensus conferences typically have leadership from North America and Northern Europe.

lawyers) are often also included. The inclusion of laypersons is common, and provides a check on the perspective of experts, as well as a transparency to the whole process which allays concerns about secrecy that might impinge on perceived objectivity. The addition of patient representatives (sometimes in addition to, sometimes the same as, laypersons) often provides important input about patient preferences, for example the variability in patient preferences for surgery in early breast cancer. In the Danish consensus conference model, laypersons can also represent the public voice in policy deliberations, in a model of public democratic deliberation.

Different countries can productively hold consensus conferences on the same topics. For example, recommendations for mammograms to screen for breast cancer may be different in countries with different resources or different values. In the UK, mammograms are not routinely offered to elderly women because they are not assessed as cost-effective at that age; in the United States cost-effectiveness is not valued so highly. Moreover, national barriers are significant enough that representatives from a group in one country are often not seen as negotiating for representatives of the corresponding group in another country (for example, participation of obstetricians in the Netherlands in a consensus conference on cesarean delivery would probably not reassure obstetricians in the United States that their perspective had been represented on the panel).

Just as the NIH Consensus Development Conference Program declined with the rise in evidence-based medicine, so has the number of European consensus conferences—at least, those which explicitly use the term "consensus conference" for medicine. Many of the 1980s and 1990s consensus conference programs, such as those of the King's Fund Forum in the United Kingdom, the Norwegian Institute for Hospital Research, and the Norwegian National Research Council, probably no longer exist (I have been unable to find evidence of recent activities). Some have been transformed and renamed to emphasize evidence-based medicine and have dropped the description "consensus conference," although they still partly rely on face-to-face meetings and group processes. Yet group judgment is *still* a pervasive feature of medical technology evaluation. For example, the Canadian Task Force on Preventive Health Care—one of the earliest consensus programs, started in 1979—was completely revamped to emphasize "Evidence-Based Prevention" which are the words with the largest font on their brochure in the early 2000s.[22] Group judgment is still part of the process, which uses a "National scientific panel of clinician-methodologists."[23] (A similar renaming and revamping process took place in the United States a little later. For example, the Medicare Coverage Advisory Committee, established in 1998, was renamed the Medicare Evidence Development Coverage Advisory Committee in 2007.)

## 3.4   The End of Consensus Conferences?

Now that consensus conferences typically follow an evidence assessment and are often renamed, leaving out the term "consensus conference," in what sense, if any, are they still important for making medical knowledge? The NIH Program and several of the European medical consensus conference programs no longer exist. Is it just a matter of time before all consensus conferences fade away? The evidence-based medicine movement has ranked expert consensus as the lowest level of evidence, when it is counted at all. Has it struck a fatal blow to the consensus conference movement? I have two sets of responses to these questions.

My first response is that, whether or not the term "consensus conference" is used, the same basic ritual of consensus conferences continues. Face-to-face meetings lasting 2–3 days with 10–20 invited people, usually experts and major players in the field along with some public representation, are still as familiar as they were 20 years ago. Such meetings produce guidelines, recommendations, and standards of care. The groups producing these statements include professional societies and specialty organizations, public and private organizations, government agencies at the federal, state, and local level, international professional societies, and health care organizations or plans.[24] Of course, there were face-to-face meetings and committee reports in medicine before the NIH Consensus Development Conference Program, but the program gave a huge impetus

---

[22] This was available online, but is no longer (as this program was discontinued in 2005 and recreated in 2010).

[23] Note the new importance of "methodologists"—people with the skills to produce systematic evidence reports. The traditional importance of "clinicians" is not given up, however.

[24] See the summary statement at the National Guideline Clearinghouse website <http://www.guideline.gov/about/inclusion-criteria.aspx>, accessed July 20, 2014.

to these practices and a standard format for group decision making in the medical context which influenced all that followed. A tradition was established that is still in wide use.

My second response to these questions is to say that evidence-based methods are not relevant to any task of "interface consensus," which aims to get agreement on policy implications. A good "interface consensus" depends on a democratic deliberative process among stakeholders, who may be medical, political, or public representatives. I have not yet said much about how successful consensus conferences are at broader political, ethical, and policy discussion and negotiation, or to what degree they follow the democratic deliberative model. There have not been formal evaluations of this. My impression is that some of these negotiations are successful in coming to consensus on practical matters and getting buy-in from relevant constituencies. I do not know the degree to which they follow a democratic deliberative model, and it would take a sophisticated methodology, as well as access to observe "executive sessions," to discover the actual processes of negotiation. I will say more about the nature of democratic deliberation in Chapter 4.

We have some indirect evidence about the workings of "interface consensus" in the experience of the so-called "Danish consensus conference." This was designed by the Danish Board of Technology in the mid-1980s, again based on the NIH Consensus Development Conference Program and including the infamous late-night writing sessions. It is not a medical consensus conference program; it deals with other questions of technological efficacy and suitability. The most distinctive modification to the NIH process is to the composition of the panel: instead of a mixed panel of various kinds of experts and a few laypeople, the entire panel is composed of laypeople. Experts—scientific, ethical, economic, etc.—play a consulting role and not a decision-making role. Lay panelists are asked to consider the general welfare, rather than their personal preferences, in deliberations. The process is thus different from that of an unreflective and uneducated mass poll. A Danish Board of Technology employee writes that the Danish model offers a "new way of hearing 'the voice of the people'" (Andersen & Jaeger 1999: 339). The term "public enlightenment" is used. This is a Danish concept based on the *folkelig* concept of the common good, which comes from the formation of the Danish nation-state in the nineteenth century and the ideas of the poet and priest N.F.S. Grundtvig (Horst & Irwin, 2010). The model gained additional resonance through its coherence with the political philosophy of Jürgen Habermas. In Denmark, and in other European countries that adopted this model (e.g. Sweden, Norway, the Netherlands, and the United Kingdom[25]), the results are used to advise the parliamentary process. The Danish model became well known in its own right, and its origins at NIH in the 1970s were sometimes forgotten.[26]

The main disappointment with all consensus conferences—whether technical or interface—is that they have not been successful at dealing with serious controversies. In controversial cases, dissenters have been more inclined to find fault with the process (looking for instances of bias and unfairness) than to change their views. If their function is to produce consensus, then consensus conferences fail where they are most needed.

The future of the Danish consensus conference model is uncertain. The Danish Board of Technology has been experimenting with other group processes, such as Café Seminars, Citizen's Hearings, Citizen's Jury, Future Panel, Hearings for Parliament, Interdisciplinary Work Groups, the Voting Conference, and the Scenario Workshop. In late 2011 it lost its public funding. It continues as a private organization, the Danish Board of Technology Foundation.

Meanwhile, in the United States the idea of using consensus conferences to model public participation regularly brings up the concern that the population is too diverse to be modeled by 10–20 persons having a 2–3 day discussion. There are some attempts, so far not very successful, to move to virtual deliberation, which can include more people (Delborne et al. 2009). I watch these with interest because I wonder about the dependence of the consensus conference model on good old fashioned face-to-face communication of a group of people countable on fingers and toes. It is possible that trust requires this traditional form of human interaction, but before coming to such a conservative conclusion it is worth experimenting with what can be done in moving from face-to-face to keyboard-to-keyboard, and from 10–20 persons to indefinitely large groups.

[25] Richard Worthington claims that the Danish model has been "tested or adopted in at least 42 nations on every continent" (2011).

[26] A 1999 paper by David Guston is titled "Evaluating the first US consensus conference: the impact of the citizens' panel on telecommunications and the future of democracy". The paper is on a conference held in April 1997.

## 3.5  Disagreement over Consensuses and Guidelines

There is often more than one consensus conference and/or set of guidelines on particular medical topics. When consensuses and/or guidelines differ, confidence in the conclusions is undermined, unless it can be shown that the differences are due to differences in interface considerations. Thus if two different sets of guidelines are produced in the same country for the same medical technology, it will look like consensus conferences depend on the composition or conduct of the panel in a contingent and arbitrary fashion (unless it can be shown that the differences are due to differences in political, ethical, or other interface matters).

This could be put directly to the test by holding two (or more) independent meetings on the same topic at the same time.[27] This has not been done intentionally with the major consensus development programs.[28] Several small studies, however, have looked at the influence of professional specialty on panel conclusions. There are no great surprises in these studies: groups composed of chiropractors are more likely to favor spinal manipulation for lower back pain than multispecialty physician panels (Coulter et al. 1995), surgical panels are more likely to recommend carotid endarterectomy than multispecialty panels (Leape et al. 1992), and general surgeons are more likely to recommend cholecystectomy than a mixed panel of gastroenterologists, general practitioners, surgeons, and radiologists (Scott & Black 1991). These differences occur even in cases where all panelists are given the same review of the evidence prior to deliberation (Scott & Black 1991). It is unrealistic to expect the group process to eliminate, or even reduce, such bias.[29]

External review of guidelines before publication probably reduces disagreements, provided that the review panels are sufficiently diverse. Most non-NIH programs incorporate this level of review.

---

[27] One could test the replicability of the results of one program by running two or more conferences with the same format independently, or test the consistency of the results of two different programs by running two or more conferences with different formats independently.

[28] It has happened, in an uncontrolled way and not at the same time, for recommendations on screening mammography. See Chapter 9 for a discussion of this issue.

[29] Awareness of group process, and use of strategies to avoid bias, can of course help.

## 3.6   The National Guidelines Clearinghouse

There are so many different groups producing consensus statements and clinical guidelines for treatment of so many conditions that it makes sense to collect them in a way that makes them maximally accessible. Several thousand current clinical guidelines are archived at The National Guideline Clearinghouse (NGC), run by the AHRQ. The NGC sets standards for guidelines to be listed, lists current guidelines, and produces guideline syntheses and guideline comparisons.[30]

The standards for listing guidelines are strict. They include recency (developed and/or revised within the last five years), a background evidence review including a systematic literature search, and sponsorship by a professional organization. The kind of organization is specified: "medical specialty association; relevant professional society, public or private organization, government agency at the Federal, State, or local level; or health care organization or plan."[31] Individuals and research groups cannot submit guidelines, no matter how evidence-based they are. I interpret this organizational requirement as a confirmation of the epistemic, political, and rhetorical importance of appropriate group judgment.

These days, guidelines are typically developed *in addition to* and following a systematic evidence review. For the NGC, guidelines should be evidence-based so far as is possible; they should make it clear when a recommendation does not have adequate evidence. "Expert agreement" is required *in addition to* evidence for the NGC. This concurs with what I said in Section 3.5 about the continuing need for expert group consensus even in an age of evidence-based medicine.

## 3.7   Conclusions

Medical consensus conferences are not what they were originally named and sometimes still claimed to be. They do not resolve medical controversy and bring about consensus on scientific matters. Nevertheless, they have important functions. All consensus conferences are somewhere on the scale of "technical" versus "interface" concerns, and can contribute

---

[30]  I will discuss guideline syntheses and guideline comparisons in Chapter 9.
[31]  <http://www.guideline.gov/about/inclusion-criteria.aspx>, accessed July 20, 2014.

to epistemic and practical goals. The NIH Consensus Development Conference Program is at the extreme "technical" end of the scale and the "Danish model" is at the extreme "interface" end of the scale. At their best, consensus conferences can help "close the gap between research and practice" by disseminating evidence-based results in an authoritative manner. Also at their best, consensus conferences can provide a procedure for negotiating ethical, political, and policy matters related to medical technology in a way that will prove acceptable to the medical and wider community. The important difference between the American (US and Canada) and the European contexts is that in the American context the experts negotiate, whereas in the European context there is a much greater role not only for public participation but also for public deliberation on social, political, and ethical issues that directly feeds into governmental decisions.

Consensus conferences may decrease in importance if and when clinicians and the public become more trusting of the results of evidence-based medicine and, perhaps also, more knowledgeable about the potential for bias in apparently democratic group deliberation. They may increase in importance, and perhaps evolve further, if our skill at or confidence in democratic deliberation grows.

# 4

# Philosophical Interlude
## Objectivity and Democracy in Consensus Conferences

Great minds think alike but fools seldom differ. (Anonymous)

Those who know that the consensus of many centuries has sanctioned the conception that the earth remains at rest in the middle of the heavens as its center, would, I reflected, regard it as an insane pronouncement if I made the opposite assertion that the earth moves. (Nicolaus Copernicus, Preface to *De Revolutionibus*, 1543)

Scientific controversies can be resolved only by rigorous testing and critical evaluation . . . Scientific controversies, unlike public policy decisions, cannot be resolved by consensus. (Alfred E. Harper, quoted in Gunby (1980) in a *Journal of the American Medical Association* editorial)

## 4.1 Introduction

Consensus is both rhetorically and politically powerful. So far, in Chapters 2 and 3, I have argued that medical consensus conferences use the rhetorical power of technical (scientific) consensus and the political power of interface consensus to achieve their goals of disseminating new knowledge and/or negotiating choices about a new technology.[1] My goal in this chapter is to look at both consensus and consensus conferences more deeply. Consensus conferences have developed from ordinary consensus practices and from consensus in science. It is worth paying some attention to how consensus conferences may both resemble

---

[1] Some parts of this chapter build on Solomon (2006).

and differ from ordinary consensus practices and consensus in science. Then I want to look at the two normative concepts most often associated with consensus conferences—"objectivity" and "democracy"—and ask whether either of those ideals are attained in consensus conferences. The same method—"rational deliberation"—supposedly achieves both of these ideals. Rational deliberation basically means that panelists should engage in reasoned and democratic argument with one another, giving weight to strong arguments and speakers with earned authority, and ignoring "irrelevant" facts such as the demographics of speakers or the volume of their speech. Rational deliberation is a common term from philosophical literature, especially the literature in the areas of social and political philosophy. It is also an ideal, and empirical work on group process suggests that, often enough for concern, the ideal is not achieved. The implications of this for both "technical" and "interface" consensus conferences will be discussed.

## 4.2  Consensus Practices

There were consensus practices, informal and formal, long before there were consensus conferences. Consensus decision making is informally used in many aspects of private life, within families and friendships. The alternatives to consensus building are voting (but this is not common in private life) or the exercise of power such as in patriarchy or oligarchy. Many group decisions are a mixture of democratic consensus and a differential exercise of power.

Consensus approaches can also be used in community life. For example, the Religious Society of Friends (Quakers) has used formal consensus processes since the seventeenth century, and kibbutzim (cooperative farms in Israel) have governed themselves by consensus since their origins over 100 years ago. Some non-Western societies also have traditions of consensus decision making (Gelderloos 2006). Typically, there are traditions and sometimes explicit procedures to follow for coming to a consensus decision in these different contexts.

Groups that govern themselves by actively seeking consensus are thought of as deeply—sometimes too deeply—democratic. This is because each individual is empowered to block a decision if, in their judgment, it is not in the best interest of the group (Gelderloos 2006). This can make the consensus process lengthy and difficult, and sometimes impossible. Groups that take a vote and proceed with a majority or a two-thirds vote are more typical democratic institutions, especially in the US, and in these groups individuals may vote either for the best interest of the group or for their own personal interest. Usually, only small groups (under about 20 persons) go through the sometimes lengthy procedures necessary for forming consensus.[2]

Consensus processes are more likely than majority votes to produce commitment to a plan of action in each member of the group (Dressler 2006). This is because the active consent and participation of each member is required to produce consensus. In contrast, decisions which are made by a voting majority often leave the minority disenfranchised and less likely to buy in to the plan of action. Producing a high level of *commitment* to a plan is a great advantage of consensus processes.

Coming to consensus often requires *negotiation*. Successful negotiation generates a solution—sometimes a highly creative one—that acknowledges the different interests of the involved parties yet finds a way for them to cooperate in the service of their joint and separate goals and perspectives. Negotiation is a process that requires participation (by the different interests involved), creativity (on the part of the facilitators and the participants), and flexibility (from the participants). The success of a negotiation is not a matter of finding the "correct" answer, but rather an answer that "works" in that it ends the dispute for long enough that the group can do something productive. That is, coming to consensus on some issues is a prerequisite for voluntary collective action. There may be more than one democratic solution to a negotiation, and there may be different kinds of solutions for different contexts.

## 4.3  Consensus in Science

Science is not a solitary enterprise. Epistemic trust[3] in the work of both predecessors and contemporaries is foundational, as Hardwig (1991) and others have argued. Epistemic trust can be in epistemic authorities, epistemic peers, or even epistemic inferiors, with different levels of trust for different levels of expertise and domains of inquiry. (So, for example,

---

[2] An exception is small Quaker colleges, such as Haverford College, that are governed by faculty consensus. Haverford College has approximately 100 faculty members.

[3] Epistemic trust is reliance on the truth of someone else's (or some group's) beliefs.

a senior scientist can trust a graduate student to record the data faithfully, but not to interpret it.) Before the Scientific Revolution, epistemic trust often rested on well-known ancient texts; in medicine, specifically, the Galenic corpus, which was treated as authoritative. The Scientific Revolution challenged arguments from the authority of ancient science, but did not do away with epistemic authority. Epistemic authority is now understood as an earned consequence of expertise, which is gained through training and experience. When an expert voices an opinion on a matter about which she has expertise, it is reasonable to think that the opinion is more reliable than a layperson's opinion. When experts agree with one another (expert consensus) there are even stronger reasons for trust; when experts disagree with one another (expert dissent) their authority is reduced or lost altogether. But as the introductory quotations suggest, expert consensus is not a mark of infallibility.

The quotation from Copernicus which I used as an epigraph for this chapter shows a common epistemic use of appeals to consensus in science—they are used in arguments from authority. The idea is that the experts arrived at consensus because each one of them reasoned—independently or through joint efforts—to the same, true, conclusion. Experts do not always arrive at consensus, and the suggestion is that if they do arrive at consensus, this is because they have compelling reasons such as sufficient evidence. The implication is that if you disagree with the consensus, then you should have reasons for thinking that your reasoning is better than the reasoning of numerous experts. The rhetorical effect of expert consensus, on most people, is deference to the experts.

Anyone with a modest understanding of the history of science knows, with Copernicus, that a scientific consensus can be wrong. Nevertheless, when asked to defend a scientific position it is difficult not to appeal to a scientific consensus. A striking example of this is Naomi Oreskes' paper "The scientific consensus on climate change," which describes her study of all 928 scientific papers on climate change from 1993 to 2003, concluding that "Remarkably, none of the papers disagreed with the consensus position" that there is anthropogenic climate change. She goes on to recommend that we listen to this consensus (Oreskes 2004). Oreskes is not willing—or not able—to argue for climate change in a few pages without an appeal to the consensus of experts. This is not a fault of Oreskes; rather, it is a sign that expert consensus is indispensable in argumentation even though it is not as reliable as we would like. Guarded confidence about the correctness of any particular consensus in science is probably the most justified initial attitude.

*Arriving* at consensus is different from *negotiating* consensus, at least prima facie. Scientists do not attempt to end scientific disagreement through discussion and negotiation alone. When physicists or geologists or genetic engineers disagree, they do not convene a consensus conference to resolve their disagreements, although they often plan meetings to share and critique one another's ideas. They do not regard consensus conferences as "a rapid data synthesis method" (Schersten 1990) (as the early NIH Consensus Development Conference Program did), and they do not impose a time frame or appoint a panel to end dissent. Rather, they go back to the lab or back to the field in search of additional evidence, devising experiments and observations around points of controversy. They do not even convene a consensus conference or write a consensus statement when the controversy is over. Typically, consensus in science develops in an informal process that is more extended over time and space and involves more scientists than can be involved in a consensus conference. This informal process is typically influenced by a variety of factors, including empirical successes and relationships of power (see Solomon 2001).

"Strong Programme" constructivists (such as Barry Barnes and David Bloor, Steven Shapin and Simon Schaffer, Harry Collins and Trevor Pinch) deny the claim I just made and argue that, appearances to the contrary, scientists "negotiate truth." Their suggestion is that scientists use the same processes as groups that attempt to come to consensus on practical, political, or ethical matters. Their claims were radical precisely because it does not seem as though scientists negotiate truth; it seems as though they discover it. This is not the place to evaluate the Strong Programme, just to note that according to the Strong Programme scientific consensus is no different from negotiated consensus. I disagree with the radical position, although I acknowledge that scientific evidence and reasoning are not the only constraints on scientists' decisions.

Historical accounts of scientific change typically tell narratives about successive cycles of dissent followed by consensus. I think that this cyclic narrative is sometimes exaggerated, and consensus is too easily seen as the goal of inquiry by those working in science studies. Scientists do not come to consensus for the sake of consensus, and consensus conferences are not a way in which they attempt to end dissent. If asked, they argue that since their goal is truth (or something like it), it is more telling when they find themselves agreeing with each other *without* trying to do so than when they actively seek to do so.

There is one recent case of an official statement of consensus in science which might be thought a counterexample to my claim that consensus conferences are not appropriate for science: the IPCC's statements on global warming (the fifth statement is expected to be completed in 2014), which have more than 800 authors. On closer inspection, this case turns out to confirm the view that consensus in science is not negotiated through social interaction alone, and that consensus statements in science are not for the benefit of the scientific process.[4] In fact, climate scientists contain their differences in order to present a united front to the public and to governments. Sometimes, for example, they assent to a statement that is *less* alarming about climate change than their own research suggests, because even this milder conclusion is a sufficient basis for aggressive and immediate action.

The report of the IPCC is not a resolution of a scientific controversy but a professional effort to communicate to non-scientists the best of scientific thinking on a topic, with the minimal amount of agreement needed to make firm recommendations for action. Furthermore, the report is only needed because of what is happening *outside* of science, i.e. the climate change deniers.[5] In my view, the worst (yet probably intended) effect of the pronouncements of climate change deniers is to interfere with the negotiation of joint action, by insisting that there are not sufficient shared beliefs to proceed with *any* action. The point of the IPCC report is to stop this interference and foster joint action.

The IPCC report is doubly unusual in having such a large number of authors, which is not typical for a consensus statement. I think the reason for this is to maximize the rhetorical power and international reach of the report, because of the urgency and importance of the findings. Since

[4] There has been speculation that there are some benefits for scientists from producing the IPCC statements. For example, the consensus statements have pointed out areas for future research, and this can positively influence new research. These benefits were not, however, sought in advance or used as a justification for the work of the IPCC.

[5] It would take a while to argue that climate change deniers are outside science—and it is not my topic here.

"technical" consensus is not really being negotiated, and the context is a consensus statement rather than a consensus conference, large numbers are manageable.

## 4.4  Consensus Conferences

The British physician Trisha Greenhalgh (2000: 6) has a healthy disrespect for consensus conferences, going so far as to call consensus conferences "decision making by GOBSAT" ("Good Old Boys Sat Around a Table"). In the context of an introduction to evidence-based medicine, she can afford to be somewhat cynical about the results of consensus conferences. (As I will show in Chapters 5 and 6, however, evidence-based medicine in some ways rests on a foundation of expert consensus.)

Some consensus conferences—specifically "interface" consensus conferences[6]—are formalizations of ordinary consensus practices of negotiating actions. Others, specifically "technical" (scientific) consensus conferences, are *not* formalizations of consensus practices, although they may get some of their rhetoric from consensus practices (especially the rhetoric about democracy). In the case of the NIH Consensus Development Conference Program, consensus conferences did not require negotiation of conclusions, as was discerned in Chapter 2. The NIH Consensus Development Conference Program was really about objectivity and intellectual authority, and the experts were supposed to concur on the basis of consideration of the evidence, not on the basis of any need to accommodate each other. In NIH Consensus Conferences, concurrence was not expected to develop through negotiation; the hope was that it would happen because experts had an objective discussion of all of the evidence, and each expert ended up coming to the same conclusion, either for the same reasons or for different but compatible reasons.[7] Often, consensus among researchers preceded the conference, so the task was simply to bring the panelists up to speed. NIH Consensus Conferences used the

---

[6] Recall from Chapter 2 that "interface" consensus is consensus on ethical, social, political, and economic matters and contrasted with "technical" consensus, which is consensus on scientific matters. (The distinction of course presumes the separability of these matters.)

[7] In *Social empiricism* (Solomon 2001) I argued that consensus in science is often the result of different evaluations that lead to the same choice.

rhetorical power of a group of experts to certify the objectivity of the consensus and communicate it with authority.

Even with such purely "technical" (scientific) consensus conferences as the NIH Consensus Development Conference Program, the goal of the conference is not to *develop* a scientific consensus. As already argued, in general scientists do not hold consensus conferences to settle scientific disagreement. Expert consensus in science, when it forms, does so without the help of consensus conferences. The goal of "technical" (scientific) consensus conferences is typically the *dissemination* of knowledge beyond the scientific community through an authoritative statement made by a group that is trusted because of the expertise and professional standing of its members.

"Interface" aspects of consensus conferences are much more like everyday consensus processes, since they involve coming to agreement on ethical, policy, and pragmatic matters in order to take joint action. Here, negotiation is necessary, and fairness (rather than objectivity) is the guiding value. Most medical consensus conferences have some mixture of "interface" and "technical" (scientific) goals.

Therefore, consensus conferences are not institutions of basic science, even when they are devoted to purely scientific questions. They are always designed for a wider community that is making decisions informed by science. They are a feature of applied rather than of basic research. When consensus is important for joint action, conferences or other direct ways for forming or expressing consensus are appropriate. These conferences may limit themselves to stating a scientific consensus, or they may go further and negotiate policy.

When is physician consensus on medical practice important? Variability in care for the same medical condition usually means that some are not getting important care while others may be getting unnecessary care. Lack of consensus on best care encourages insurance companies or governments to reimburse only the most basic efficacious care, and physicians to order all possible care. When physicians agree on best care practices, their unified voice creates an opportunity for improving care overall. This is because such consensus can have the function of maintaining the authority of physicians in debates about health care practice and reimbursement, leaving less discretion for insurance companies and governments. Too much consensus in science or medicine is, however, undesirable, in part because it can prematurely close scientific controversy, in part because not all social, political, and ethical decisions need be collective and in part because we may wish to experiment with different social, political, and ethical decisions in different contexts. More remarks about the greater importance of consensus in medicine (and other technologies), and its lesser importance in basic science, will come at the end of this chapter and in Chapter 9.[8]

## 4.5   Philosophical Uses of Group Judgment

Contemporary moral and political philosophers make prominent use of group judgment, often writing of "deliberative democracy," "rational deliberation," or "discursive democracy." Characteristically, the Kantian focus on *individual* reason has been supplemented or replaced by similar ideals of *group* reason. Reason and/or rationality are still essential, but groups are considered to be more reliable and objective than individuals. For example, in Rawls's "original position" a community of individuals (not a single individual) rationally deliberates from an imagined position of ignorance about social position in order to select shared principles of justice (Rawls 1971). And in Habermas's "discourse ethics," participants rationally deliberate with one another from their actual social positions, attempting to reach consensus (Habermas 1971). Joshua Cohen, Amy Gutmann, and Seyla Benhabib are other well-known contemporary political philosophers who endorse deliberative democracy.

Some philosophers of science such as Mill, Popper, and Longino have, similarly, made critical discussion and deliberation central to their social epistemologies. Longino (1990, 2002) even goes so far as to claim that objectivity is constituted by such critical discourse, provided that the discourse satisfies constraints such as tempered equality of intellectual authority, public forums for criticism, responsiveness to criticism, and some shared standards of evaluation. She adds that scientific objectivity, in particular, requires sharing the value of empirical adequacy[9]

---

[8] I do not intend to make a *naïve* distinction between basic (sometimes called "pure") and applied science here. (I am not, for example, depending on the traditional view that basic science is value free.) Basic and applied science may be distinguished by the presence of different values. However, this is not the place to offer or defend a theory about the difference between basic and applied science.

[9] Empirical adequacy of a theory is due to its predictive and explanatory successes.

(1990). The general idea is that deliberation should be rigorous and robust enough to identify and eliminate individual bias and error and include sufficiently different points of view to identify assumptions and include all relevant data and considerations.

There is some prima facie plausibility in the idea that groups can be more objective than individuals. Ideally, all considerations can be shared and considered by members of the group. This overall assessment seems more thorough than the limited considerations one individual can bring to a question, more error free than the fallible conclusions of a single person, and more likely to take advantage of social and other kinds of diversity (educational, temperamental, ideological, etc.) of group members. Often, the desired group discussion is described as "rational deliberation" in which criticisms and responses to criticism are thought of as intelligible to each reflective individual. Deliberation corrects errors, uncovers presuppositions, and sometimes transmits additional evidence or reflection from one or two members to the rest of the group. Each individual in the community is thought of as capable of recognizing their own errors and presuppositions when they are pointed out by others, and of accepting evidence discovered by others. In this model of group deliberation, individuals may start out from different positions but, hopefully, all end up agreeing on the merits of and problems with each theory and sometimes even end up forming consensus on one (the best) theory. In this way, groups can achieve objectivity via the individual objectivity of each of their participants.[10] This result is more individualistic than a model of distributed judgment in which individuals may end up in the same place, but typically not for the same reasons or causes.[11]

There is also prima facie plausibility in the idea that democracy contributes to objectivity. Each person should be able to share their experiences and concerns, and have them taken seriously. In this way the group will make the best use of its epistemic resources.[12] Miranda

[10]    Helen Longino's position is a little more subtle than this. She holds that while criticisms and responses to criticism should be taken seriously by everyone, the result will not be choice of the same theory because of the diversity of values held by different communities of scientists. She does, however, conclude that the result of group deliberation is greater objectivity.

[11]    My account of scientific consensus in (Solomon 2001) is an example of a distributed model.

[12]    This is a reasonable argument, but not a conclusive one. Non-democratic groups may have epistemic advantages that outweigh this benefit of democratic groups. The connection between democracy and objectivity is, in my opinion, contingent.

Fricker (2007) conceptualizes this as what she calls "testimonial jus-tice." Democracy (fairness) is also valued for its own sake, such as when a group negotiates an "interface" consensus.

In these philosophical writings, group judgment is thought of as capable of satisfying two important constraints, one epistemic and one political: *objectivity (or rationality)* and *democracy (or fairness)*.[13] Group rational deliberation is regarded as the proper response to both intellec-tual controversy and political conflict. Consensus is desirable in applied epistemic contexts where consensus is typically regarded as a marker of objectivity, rationality, or even truth. Consensus is also the desired out-come in political contexts, where it is regarded as the result of a successful and hopefully fair negotiation in which everyone is sufficiently satisfied to assent to the conclusion. Objectivity and democracy are constituted by the *process* and not by the *outcome*, so long as the outcome is consensus.

The concept of rational deliberation in a group context is—perhaps surprisingly—a very "internalist" idea, in a sense of the word internalist that is widely used among analytic epistemologists.[14] Roderick Chisholm characterizes internalism this way: "If a person S is internally justified in believing a certain thing, then this may be something he can know just by reflecting upon his own state of mind" (1989: 7). The core of the posi-tion is that justifications for belief are internal to each individual knower and, typically, not only internal but also available to reflection. In a group deliberative context, each individual comes to understand for themselves the reasons for a belief.

The contrast is with "externalism," according to which one can be justi-fied, or have knowledge, without being able to say *why* one is justified or has knowledge. Thus externalists claim that the reasons for justification or knowledge are not (or not readily) available to individual cognizers, either because the reasons do not enter conscious awareness or because the reasons (or causes) are external to individual minds. Externalists sometimes dispense with the concept of objectivity, or reconceive it in less

---

[13] Feminist empiricists (Nelson 1997, Anderson 2004, Clough 2004) often argue that there is no difference between epistemic and political constraints. I have argued against this in Solomon (2012). Nothing here hinges on the outcome of this discussion.

[14] The concept of group rational deliberation is also (again surprisingly) somewhat individualistic, in that it depends on the rationality of *each* member of the group during and following group discussion.

intuitive or less internalist terms. For example, in Alvin Goldman's (1992, 1999, 2002) work, "reliable process" can replace traditional objectivity.

It might be thought that social epistemology is *all* externalist, since by definition social epistemology concerns itself with epistemic processes that go beyond the individual. However, ironically, this area of work—looking at deliberation and criticism—in social epistemology is internalist in character, because the epistemic processes discussed (correction of errors in reasoning, sharing new arguments, and so forth) are available to reflection and because group epistemic assessment is conceptualized as the sum of reflection by each and all of the individuals involved. Again, the contrast is with externalism, in which epistemic assessments need not be based on reflection or introspection at all, and need not be conceptualized as the sum of individual judgments. There are other methods of aggregating judgments (Solomon 2001, List 2005).

Work in social epistemology by philosophers has not always been "internalist" in the way described above. Certainly Alvin Goldman is clear in his position that social practices are to be evaluated instrumentally for their conduciveness to truth, rather than intrinsically for their rational cogency. For example, Goldman (2002) claims that individuals get their first beliefs through a process he calls "BABA": the bonding approach to belief acceptability. Goldman suggests that beliefs are contagious when there is a prior bond of affection between individuals. It turns out that BABA is reliable, and the reliability of BABA is understood through evolutionary psychology: those with affective bonds will have less incentive to deceive one another than those without affective bonds. The epistemics of BABA are not transparent to reflection.

In Solomon (2001) I used an externalist account of "scientific rationality" to make methodological recommendations in social epistemology of science. For example, I argued that it is both unnecessary and unrealistic to expect individual scientists to be free of cognitive and motivational bias. I argued that striving for the equitable distribution of bias is more important than making efforts to eliminate it.

Most philosophical discussion of social epistemology, however, is of the traditional "internalist" kind. Even the work of Helen Longino, which is nontraditional in important ways, rests on what she calls "an intuitive distinction between knowledge and opinion that I take to be shared" (2002: 174). She is saying that her four criteria of objectivity (which are

connected to her concept of knowledge)—tempered equality of intellectual authority, shared values, public forums for criticism, and responsiveness to criticism[15]—rest at least in part on intuitive plausibility.

Epistemic appeal (objectivity) and political appeal (democracy and
fairness) are not the same virtues in the scientific domain,[16] although
they may be conflated with one another rhetorically, since both use normative language—the language of moral imperatives—to achieve their
goals. Both "objective" and "democratic" are powerful words of praise in
our Western industrialized culture. Yet political fairness is relevant *epistemically* only when knowledge reached democratically is more likely to
be true (or approximately true or useful or objective) than knowledge
produced in non-democratic social contexts. There is no general or a
priori argument for this, although it has some empirical plausibility, at
least in broad outlines. Certainly philosophers such as Karl Popper made
much of the dependence of scientific progress on a democratic political
environment.

It should also be noted that the concepts "objectivity" and "democracy" are somewhat variable in meaning, and care needs to be taken to
avoid equivocation when using them. "Objective" can mean "unbiased,"
or "reliable," or "following a standard process," or it can have other
meanings.[17] "Democracy" can mean a variety of participatory and/or
representative practices.

## 4.6  Groupthink and Other Biasing Phenomena: Problems with Objectivity

Having said a good deal about the appeal of group judgment, I now turn
to some reasons for being skeptical about its results. Here are four surprising empirical results about group deliberation that have implications
for the epistemology of consensus conferences:

1. Group deliberation often produces worse decisions than can be
   obtained without deliberation. (Often enough for epistemic concern.) (Janis 1982, Surowiecki 2004).

---

[15] See Longino (2002).    [16] See note 10, this chapter.
[17] The term "objectivity" has been reappropriated by feminist epistemologists such as
Sandra Harding and Donna Haraway, who use it in novel ways.

2. A group of non-experts often produces better decisions on a topic than does an expert, sometimes even a group of experts, about that topic. (Thus challenging the traditional deference to experts.) (Surowiecki 2004).

3. If group deliberation does take place, outcomes are better when members of the group are strangers, rather than colleagues or friends. (More interpersonal understanding does not help.) (Sunstein 2003).

4. Groups and organizations sometimes don't know what they know. Sometimes groups do not put together the relevant knowledge of all the individuals in them (Surowiecki 2004). This can happen when a dissenting view is not explored (Sunstein 2003).

Group deliberation is fallible. Internalist standards for group deliberation such as sharing data and critiquing and responding to arguments may seem intuitively plausible, but in practice (in the "real world") satisfying them is neither necessary nor sufficient for good epistemic outcomes. James Surowiecki (2004) argues that, under typical conditions, aggregating opinions without deliberation produces better results than allowing groups to deliberate and come to a conclusion together.[18] This result sometimes holds even when participants are not experts. An example is the Iowa Electronic Market, which aggregates the predictions of ordinary individuals about upcoming elections, and does better than all other known methods of prediction of election outcomes. So a group of non-experts who do not deliberate can do better than a group of experts sharing insights and deliberating with one another. Another example is that investment groups where the members are strangers to one another do better than investment groups in which the members are well known to one another (Sunstein 2003). A final example is the case of United States' intelligence before 9/11, which had more relevant information about the pending terrorist attack than it made use of.

The culprits in group deliberation are groupthink, peer pressure, anchoring phenomena, and other biasing processes, which affect reasoning despite our best efforts to avoid them.[19] The phenomenon

---

[18] Arrow's Impossibility Theorem is often presented as the main problem with group deliberation. But this theorem applies to systems of voting, not to methods of coming to consensus.

[19] One could argue that, strictly speaking, if there is groupthink or other biasing processes then there is not real (rational) deliberation. But then rational deliberation may be an unrealistic ideal.

of groupthink was first studied by Irving Janis (1982, 1972) in political contexts. Janis argued that under conditions such as high social cohesiveness, directive leadership, individual homogeneity, and isolation from external influences, groups are likely to come to incorrect or defective conclusions. Peer pressure, as well as pressure from those in authority (if present in the group) leads dissenting individuals to change their minds and, significantly, not to share their knowledge of contrary evidence. The dynamics of groupthink tend to lead the group to a polarized position, not to an average or a neutral aggregate of individual opinions. Psychological processes that lead to groupthink and other kinds of poor group decisions include anchoring (in which the first person who speaks biases the discussion toward their own views), peer pressure, confirmation bias (which leads to group polarization), and authority biases. Often-cited examples of groupthink include Kennedy's "Bay of Pigs" fiasco and NASA's decision to launch the *Challenger* spacecraft.

To take the "Bay of Pigs" example (in which the Kennedy administration attempted to overthrow the Castro government by secretly returning a small brigade of Cuban exiles to Cuba), Janis rejects the official explanation, which was in terms of right-wing political pressure, difficulties of a new presidency, secretiveness about the planning, and individual fears about reputations. Or at least, Janis claims that the official explanation is not enough to explain what actually happened: a process in which knowledgeable and intelligent advisors succumbed to the dual illusions of unanimity and invulnerability, suppressing both their own doubts and the doubts of others and acting with docility under Kennedy's leadership (Janis 1982).

While there is some controversy about whether groupthink phenomena occur in exactly the way that Irving Janis describes, there is little doubt that factors such as peer pressure, pressure from authorities, pressure to reach consensus, time pressure, and the presence of particularly vocal group members who may anchor decisions can lead groups into making poor decisions. It doesn't help much if individuals "try harder" to be unbiased or independent; we are often unaware of, and/or largely unable to resist, the social psychological factors causing groupthink and related phenomena. Solomon Asch and Stanley Milgram's classic experiments in the 1950s and 1960s show the surprising strength of pressures to conform. Kahneman and Tversky's classic work in the 1970s (Kahneman et al. 1982) shows the power of cognitive biases such as salience, anchoring, and belief perseverance. As James Surowiecki (2004: xix) puts it, "too much communication, paradoxically, can actually make the group as a whole less intelligent."

This means that the results of group deliberation can be doubtful, even when group members try their best to be objective. The result can be better when the group follows a *structured procedure* designed to avoid groupthink and other biasing phenomena. There are a number of suggestions about what such a structured procedure should be. The Delphi technique, which begins with experts anonymously sharing their initial opinions and then reconsidering and sharing again, in several iterations (until consensus is reached) is one such method, which aims to prevent both anchoring and undue influence by powerful individuals. Janis himself recommends non-directive group leaders, encouragement of (rather than mere toleration of) dissent, and several independent groups or subgroups charged with the same problem, in order to see whether or not they come up with the same result. Surowiecki (2004) recommends diversity of group membership, claiming that it brings different perspectives to a group, and also aggregation of opinions rather than attempts to come to consensus. Sunstein (2003) recommends both diversity and active encouragement of dissent. Sunstein also goes further and explores the kind of diversity that is relevant to particular problem domains. Epistemically useful diversity is not the same as social diversity, although it may coincide with social diversity for some problem domains. For example, race and gender diversity of scientists is epistemically advantageous in social scientific work on race and gender. Less obviously, Keller (1983) and others have argued that gender diversity contributes to intellectual diversity in both the biological and the physical sciences.

These suggestions are all reasonable, but they have not been formally tested and evaluated. Even Janis expresses the concern that suggestions for improving group decision making will have "undesirable side effects" (1982: 252). Some leaders, for example, might be unable to tolerate dissent without subtle or overt disapproval, which will then squash dissent. Or some leaders, for example, might be incapable of neutrality, especially when a group is coming to consensus on a view that the leader disagrees with. Or, for example, criticism and dissent might produce anger and demoralization that distracts the group from its epistemic goals. So it is not enough to make plausible-sounding suggestions for improvement. It is important to show that the suggestions can be implemented and that they lead to the desired results without overwhelmingly negative side effects. The experimental work is not there yet.

I see the variety of procedures in consensus conferences (as discussed in Chapters 2 and 3) as experiments with different types of group deliberation for different intended audiences. Audiences that are sophisticated about the epistemics of group processes (primarily some academic audiences) demand more than audiences composed of health care consumers and clinicians (who are typically more naïve about group process), in order to trust the results. For example, the NIH Consensus Development Conference Program has resisted recommendations to use structured deliberation (see Chapter 2), perhaps because this might be perceived, by naïve audiences, as constraining an open process. On the other hand, MedCAC, which reports to a more sophisticated audience, uses a more structured group process (see Chapter 3).

## 4.7   Problems with Democracy

There is more to say about what makes a conference democratic and fair than a list of who sits at the table. Society and its relationships of power do not disappear when a diverse subset engages in negotiations. As Nancy Fraser (1990), Jane Mansbridge (1990), Iris Young (2001), and others have argued, what looks like rational deliberation can in fact be "a mask for domination" in which disadvantaged groups are marginalized in discussions. Moreover, some of the same dynamics that are responsible for groupthink or other biased outcomes in technical consensus conferences can also lead to undemocratic conclusions in interface consensus. So far as I know, the conduct of medical consensus conferences has not been shaped by more nuanced views of democratic deliberation. This may be in part because "interface consensus" has been most successful in European countries (especially Denmark) in which populations are relatively homogeneous and egalitarian. Also, the *appearance* of objectivity and fairness is what matters, even though the reality of a fair deliberation may be difficult or impossible, and most people do not have a sophisticated understanding of the role of power in group discussion.

## 4.8   An Alternative: Aggregated Judgments and Preferences

Group judgment should be distinguished from *aggregated* judgments, in which members of a group typically do not deliberate with each other, but instead cast their votes independently and let them be aggregated mathematically (usually there are several choices for how to do this aggregation[20]). Financial markets such as the Iowa Electronic Market make such aggregated judgments, often with high predictive success. Jan Surowiecki's book *The wisdom of crowds* (2004) generally favors aggregated judgment over group judgments that incorporate group process. Only one medical consensus conference program that I am aware of—MedCAC, discussed in Chapter 3—aggregates judgments, and even here there is group discussion before voting so the votes are not fully independent. The medical consensus conference approach is that of group deliberation to consensus and not aggregation of individual judgments.[21]

Perhaps an aggregation technique would be more "objective," in that it might lead to better (more accurate/true) judgments, at least in some circumstances. Aggregation is decision making without discussion or negotiation. However, we (especially ordinary people who are unsophisticated about group process) are understandably reluctant to give up the intuitively plausible approach of sharing information and group deliberation.

An aggregation technique can also be more "democratic," in that each person's contribution has the same weight[22] and participants are prevented from exerting power over one another's decisions because of the absence of discussion. But such a democratic process is unlikely to result

---

[20]   See the work of Christian List (e.g. List 2005).

[21]   John R. Clarke has suggested to me in conversation that if groups aggregate decisions without prior discussion, their members may not do adequate preparation beforehand. So he is suggesting that a benefit of group deliberation is that those who expect to participate in it will do their homework.

[22]   There are aggregation techniques in which some people's judgments (e.g. experts) can be weighed more heavily than others, but they are not of interest here.

in consensus. To the extent that the purpose of coming to consensus is practical ("interface" consensus) we need the buy-in of all the participants in the process.

## 4.9   Consensus Conferences as Rituals

Some aspects of medical consensus conferences seem arbitrary or even strange, but are rarely commented upon. Here are just three observations. There seems to be a magic number of 10–20 persons on the panel for consensus conferences (the number of fingers and toes?). It seems to be necessary for consensus panels to meet "face to face."[23] Most consensus panels have a "closed-door session" which is not open to the public and not recorded for later analysis. The functions of these features of consensus conferences are not obvious, although we might hypothesize functions: perhaps 10–20 persons is the maximum size of a group in which effective group discussion and negotiation can take place, perhaps "face to face" is necessary for trust and negotiation, and perhaps "closed-door sessions" are needed in order for panelists to speak freely. But these functions are all "perhaps." In fact, we have not seriously attempted other group methods.

With the popularity of evolutionary psychology, it is tempting to speculate that some or all of the choreography of consensus conferences is simply human, or even more broadly primate, group behavior, and as basic as shaking hands at the beginning of a professional meeting.[24] We can speculate that 10–20 persons is about the size needed for group cohesion and/or trust. Closed-door sessions may make the workings of the group *less* transparent, which may *add* to their authority (in the same way that it is best not to see sausages being made). And, as Harry Marks (1997: 226) has suggested, closed doors may make negotiations easier because "concessions are easier to make when they are not publicly acknowledged as such."

These are all speculations. We do not know the degree to which current constraints on consensus conferences are historical (cultural), biological, or task oriented. We do not know the ways in which current

---

[23] Guston (1999) has written about the difficulties, so far, with K2K (keyboard to keyboard) approaches.

[24] Shaking hands probably has both animal and cultural origins.

constraints are contingent or arbitrary and perhaps also ritualistic. Moreover, there is probably an interaction between such natural, practical, and cultural elements. The moral that I take from these reflections is that this is not the time to conclude that any consensus conference procedures are necessary; this is the time to experiment with different formats, while acknowledging the depth of tradition behind the usual format. This is exactly what the Danish Board of Technology (now the Danish Board of Technology Foundation) has been doing for the last 15 years in experimenting with a variety of participatory models such as Café Seminars, Citizen Hearings, Citizen Juries, Citizen Summits, Future Panels, Interview Meetings, and Voting Conferences (see Chapter 3).

## 4.10  Standardization

Timmermans and Berg (2003) have noticed another feature of consensus conferences, which is that they can lead to standardization of medical procedures.[25] Often they are explicitly intended to do so.[26] Timmermans and Berg see this standardization as politically important because it is an indicator of professional authority. Those professions with their own standards, which they enforce, are less likely to be interfered with by other organizations with their own preferences and priorities—in this case, insurance companies, governments, and other regulatory bodies. The downside to standardization is the loss in autonomy for individual professionals. There is a tension between what is needed to maintain physician authority (i.e. unanimity) and the exercise of physician autonomy (which permits differences in professional judgment). This topic will be taken up again in Chapter 8.

In their early years, consensus conferences bore the brunt of the criticism that their results interfere with physician autonomy, and/or with

---

[25] Standardization is desirable in medicine when there is a best choice of treatment for a clearly defined patient group. It is not so desirable when the evidence for treatments is in flux, or when there is considerable known variability in the suitability of the intervention for the patient population.

[26] The NIH Consensus Development Conference Program tries to avoid "dictating the practice of medicine," but other consensus conferences, especially those run by professional organizations, are more willing to do so. Even then, soft imperatives such as "guidelines" (rather than "algorithms" or "standards" for example) often describe the result.

the "art" of medicine. These days, evidence-based medicine is first in line for such criticisms of being "cookbook medicine." In fact, I sense that consensus conferences look more personal, as well as more authoritative, than evidence syntheses, at least to non-researchers. In this culture, group of 10–20 named experts may have more rhetorical force than the more anonymous product of an Evidence-Based Practice Center.

## 4.11  Conclusions

Rational deliberation to consensus is a powerful ideal, promising both objectivity and democracy. Consensus conferences attempted to formalize the process of rational deliberation to consensus and to use it both for scientific questions and political negotiations. But there are good reasons for concern that unstructured group process (informal and formal) does not live up to its ideals, because of groupthink, testimonial injustice, and other biasing factors.[27] These reasons, as well as the importance of having scientific dissent, argue against holding consensus conferences on scientific questions when there is dissent in the scientific community. Expediting scientific consensus is not, in itself, helpful for scientific inquiry. Scientific consensus conferences can still have the important role of disseminating knowledge with authority, which they do in the IPCC and in NIH Consensus Development Conference Programs. Interface consensus conferences are a more familiar kind of consensus on political, ethical, and social matters, and much of contemporary social theory favors their ideal process of rational deliberation. However, there are reasons for concern about the actual rationality and democracy of group deliberation in practice. It is a good idea to experiment with different kinds of group and group process to see which work best in specific contexts.

---

[27]  Structured group process may eventually do better, but we do not yet know enough about how to do it well.

# 5

# Evidence-Based Medicine as Empiric Medicine

A new paradigm for medical practice is emerging. Evidence-based medicine de-emphasizes intuition, unsystematic clinical experience and pathophysiological rationale as sufficient grounds for clinical decision making and stresses the examination of evidence from clinical research. Evidence-based medicine requires new skills of the physician, including efficient literature searching and the application of formal rules of evidence evaluating the clinical literature. (Evidence-Based Medicine Working Group 1992)

## 5.1 A Brief History of Evidence-Based Medicine

The birth of evidence-based medicine is often identified with the 1992 *Journal of the American Medical Association* paper by Gordon Guyatt and his associates at McMaster University, quoted from above.[1] This paper indeed marks a turning point when evidence-based medicine extended from its origins in clinical epidemiology and came to the attention of the wider medical community.[2] The groundwork for evidence-based

---

[1] This chapter builds on some of the themes of Solomon (2011b) and there are occasional brief overlaps in text.

[2] The use of the term "paradigm" in the quoted excerpt from the 1992 paper is worth a brief comment. To philosophers of science, the word "paradigm" evokes Thomas Kuhn (1962). Evidence-based medicine is in some ways like a Kuhnian paradigm (Solomon 2011b), and the word "paradigm" entered common usage due to Kuhn's work. However, there is no reason to think that early evidence-based medicine researchers had specifically Kuhn's meanings of the term in mind. In e-mail correspondence with me (January 3, 2010) Kenneth Bond stated that "no clinical epidemiological texts or EBM [evidence-based medicine] texts make an explicit connection with Kuhn." I think this is correct, but I have not done an exhaustive search.

medicine was laid earlier in the twentieth century, with the development of statistical techniques, the design of the randomized controlled trial, and the beginning of the information age.

Evidence of effectiveness has always been a basis (although not the only basis) for medical practice. The term "evidence-based medicine" evokes no surprise—indeed, it suggests business as usual—and is thus often baffling to newcomers. I think that it would have been more precise[3] to have used a term such as "evidence-hierarchy medicine," to signal that in evidence-based medicine evidence is evaluated for both its quantity and its quality, with some kinds of evidence and some quantities of evidence regarded as better than others. The mathematical and experimental techniques for producing and evaluating evidence of high quality and quantity developed in the early twentieth century, especially due to the work of R.A. Fisher in agricultural contexts (Marks 1997). In the immediate post-World War II period, these techniques were adopted in medicine. A. Bradford Hill's work on the use of streptomycin for tuberculosis treatment (Hill 1952) is generally cited as the first double-masked randomized controlled trial in medicine. There were a variety of successes, hurdles, and failures as researchers attempted to bring the randomized controlled trial into general use in medical research. Among the successes were the polio vaccine field trial of 1954 and the 1955 evaluation of treatments for rheumatic fever (Meldrum 1998, Meldrum 2000). By 1970 the randomized controlled trial was a widely accepted "gold standard" of evidence in medicine (Marks 1997, Meldrum 2000, Timmermans & Berg 2003).

Randomized controlled trial methodology is typically defended on statistical grounds as being the *least likely*—all other things being equal—to yield biased[4] results. So long as allocations are double masked, the randomized controlled trial can (it is claimed) distinguish placebo from specific medication effects, avoid confounding due to selection biases, and escape biases of salience, availability, and confirmation. Other kinds of trial do not avoid all these possible biases, and so are in theory less reliable. For example, observational trials, which are widely used in medicine when ethical or practical reasons prevent randomization, are susceptible to selection biases, because those receiving the

---

[3] More precise, but also more vulnerable to objections.

[4] In Solomon (2001) I avoided the terminology of "bias." I am using it here in order to engage with the literature about evidence-based medicine on its own terms.

intervention may have systematic differences with those not receiving the intervention.

Archibald (Archie) Cochrane (1909–88) was one of the early champions of randomized controlled trials. During his time as a medical officer in a prisoner of war camp in World War II, he realized the need for a stronger evidence base for most medical interventions. After further training in epidemiology, he conducted groundbreaking studies on the epidemiology of pneumoconiosis (black lung disease) and published a short and influential book, *Effectiveness and efficiency: random reflections on health services* (1972).[5] The book was a manifesto calling for both the widespread use of randomized controlled trials and for *accessible databases of results*. He recognized, at a crucial moment, that accessibility of the evidence is just as important as the production of evidence. Cochrane is thought of as a founding grandfather[6] of the evidence-based medicine movement, because his work inspired many, including Iain Chalmers (an obstetrician), who began the Oxford Database of Perinatal Trials. This was the first attempt to do a systematic (and, at that time, pre-electronic) literature search for clinical trials on a topic. Chalmers also used a systematic approach to reviewing the research, and this led to the two volume *Effective care in pregnancy and childbirth* (Chalmers et al. 1989). Chalmers then created and led the Cochrane Collaboration, which began in 1993. It was named in Cochrane's honor shortly after his death.

The Cochrane Collaboration is an ongoing international effort (now with 27,000 volunteer contributors and over 4,000 current systematic reviews) to make regularly updated systematic evaluation of clinical data accessible to clinicians. The Cochrane Collaboration is the leading organization doing systematic reviews of evidence for medical interventions. Many countries also have their own organizations: for example, the AHRQ currently funds 13 evidence-based practice centers in North America which produce systematic reviews.

Evidence does not only need to be produced and recorded in an accessible manner; it also needs to be assessed for its strength. When there is

    [5] Available at <http://www.nuffieldtrust.org.uk/sites/files/nuffield/publication/Effectiveness_and_Efficiency.pdf>, accessed October 17, 2014.
    [6] Calling Archie Cochrane the "founding grandfather" allows Gordon Guyatt, David Sackett, Brian Haynes, David Eddy, and Iain Chalmers to be "founding fathers," as is appropriate.

more than one clinical trial relevant to a particular clinical question, and especially when different trials of the same intervention produce different results, we need an overall (or aggregate) evidence assessment. This is exactly what the NIH Consensus Development Conference Program hoped to achieve through expert group deliberation. There were no rules for evidence synthesis in the 1970s and 1980s. NIH Consensus Conferences made "intuitive" (or "seat of the pants") judgments in an unstructured group process.[7] As discussed in Chapter 2, however, these early consensus conferences did not perform this role transparently, or, when there was controversy, convincingly. Daly (2005: 76–7) reports a story that the evidence hierarchy was devised by David Sackett in the context of a consensus conference in which it was difficult to reach agreement because panelists each favored their own clinical experience. Sackett reportedly insisted that the entire evidence base, including clinical experience, be graded for quality. The resulting structured decision, with clinical experience low on the evidence hierarchy, relied more on the evidence for randomized controlled trials. Whether or not the story is true, it shows the perceived potential for an evidence hierarchy to resolve controversy when group deliberation fails. These days, evidence-based medicine has developed to the point that "intuitive" group assessments of complex evidence no longer have credibility.[8]

At the same time as Iain Chalmers was working at Oxford on *Effective care in pregnancy and childbirth*, epidemiologists at McMaster University in Canada (especially David Sackett, Gordon Guyatt, and Brian Haynes) were developing techniques of meta-analysis and systematic evidence review. Both techniques make use of hierarchies of evidence. Meta-analysis combines the data sets of the highest-quality studies. Systematic evidence review aggregates the results (rather than the data) of a group of studies of varying quality, producing an overall recommendation. The McMaster group coined the term "evidence-based medicine," and joined forces with Iain Chalmers in establishing the Cochrane Collaboration. David Sackett moved to Oxford in 1994. Evidence-based medicine was embraced in Canada and the UK in the

---

[7] The potential for bias in this process was discussed in Chapter 4.

[8] As described in Chapters 2–4, more recently most consensus conferences have incorporated an evidence synthesis into the preparations for the conference, thus no longer leaving evidence assessment to group process.

1990s, and adopted in many other countries.[9] Reception in the US was particularly mixed,[10] although there was strong support in some sectors, e.g. Thomas C. Chalmers (no relation to Iain Chalmers) at Mount Sinai and Dartmouth, Donald Berwick at Harvard, David Eddy at Stanford and Duke, and the medical decision making community (about which I will say more in Section 5.3).

So, while the consensus conference movement originated in the US, the evidence-based medicine movement began in Canada and the United Kingdom. These national origins and contexts shaped the movements and contributed to their successes and failures.

Much effort has gone into trying to standardize both clinical trial methodology and systematic evidence review. Since 1993, the CONSORT (Consolidated Standards of Reporting Trials) Statement, which is frequently updated, has provided a shared framework for standardizing methodology, reporting results, and assessing the quality of clinical trials for over 600 journals and editorial groups worldwide.[11] It has proved more difficult to standardize systematic review, and several different hierarchies are in use. The GRADE (Grading of Recommendations, Assessment, Development, and Evaluation) Working Group is the most prominent attempt to standardize the evidence hierarchy. A helpful history of systematic review is given in Bohlin (2012). I will say more about these attempts at standardization in Section 5.2.

There is a vast literature about evidence-based medicine—especially vast compared with the modest literature on consensus conferences. This is perhaps because evidence-based medicine has had a much more substantive effect on medical research and practice, requiring changes at every level. Consensus conferences were an addition, not a change, to usual practice. A portion of the literature about evidence-based medicine is a critical engagement with the methodology as a whole, pointing out strengths, developing advances, and discussing difficulties and limitations. This critical engagement literature is from outsiders as well as insiders to the field of evidence-based medicine, including philosophers of science and clinicians as well as epidemiologists and methodologists.

---

[9] Daly (2005) gives a useful Anglophone international history of the evidence-based medicine movement.

[10] Reasons for this will be examined in Chapter 8.

[11] See <http://www.consort-statement.org>, accessed June 26, 2014.

There is a remarkable consistency in this literature; the same concerns are raised repeatedly.[12]

In the early 1990s, evidence-based medicine advocates stressed the importance of producing and evaluating highest-quality evidence, and glossed over the skills required to put evidence-based medicine into clinical practice in the care of individual patients. Criticism that this is "cookbook medicine"[13] that devalues the importance of clinical judgment led to a quick reformulation of the basic statement of evidence-based medicine. Sackett's often cited and carefully crafted 1996 redefinition is, "Evidence-based medicine is the conscientious, explicit and judicious use of current best evidence in making decisions about the care of individual patients" (Sackett et al. 1996: 71). In general, evidence-based medicine has moderated its position and relaxed the evidence hierarchy somewhat. My discussion of evidence-based medicine will acknowledge the development and moderation of the position over time.[14]

My goal in this chapter, Chapter 6, and Chapter 8 is to develop and evaluate the criticisms of evidence-based medicine. Some criticisms seem more cogent than others, at least at first glance; I will also take the time to explore what lies behind those criticisms that seem less cogent. In this chapter, I will focus on just one type of (cogent) criticism: the claim that evidence-based medicine ignores, or devalues, the role of the basic sciences in medical research and practice. Sometimes this is expressed as the claim that evidence-based medicine devalues "psychophysiological reasoning" and/or "mechanistic reasoning." Occasionally, and intriguingly, it is described more positively as "the beauty" of evidence-based medicine "that it seems to operate at a level of scientific theory autonomous from the basic sciences" (Ashcroft 2004: 134). The observation that evidence-based medicine seems to ignore the basic sciences, while not quite correct, is particularly instructive, because it reveals the core methodological characteristics of evidence-based medicine. In Chapter 6

---

[12]  I do not manage to cite everyone who raises particular issues (for this I apologize in advance), but I hope that I have covered the content of all the important issues. My discussion is arranged topically, rather than by taking each critic separately.

[13]  "Cookbook medicine" is/was also used as a criticism of the clinical guidelines produced by consensus conferences. Whenever general recommendations are made, the "cookbook" criticism can be raised. See Section 2.4.

[14]  An excellent account of the moderation of the evidence-based medicine movement over time is given in Bluhm & Borgerson (2011).

I will look at a range of criticisms about evidence-based medicine's internal and external validity, and its reliability. That will leave a set of concerns that evidence-based medicine leaves out the "art" of medicine or the "humanism" of medicine, and I will look at these more closely in Chapter 8.

## 5.2  Techniques of Evidence-Based Medicine

The techniques of randomized controlled trials and other clinical trials, meta-analysis, and systematic review have produced an extensive and powerful body of research. A canonical and helpful definition of evidence-based medicine[15] is that of Davidoff et al. (1995: 1085–6) in an editorial in the *British Medical Journal*:

In essence, evidence based medicine is rooted in five linked ideas: firstly, clinical decisions should be based on the best available scientific evidence; secondly the clinical problem—rather than habits or protocols—should determine the type of evidence to be sought; thirdly, identifying the best evidence means using epidemiological and biostatistical ways of thinking; fourthly, conclusions derived from identifying and critically appraising evidence are useful only if put into action in managing patients or making health care decisions; and finally, performance should be constantly evaluated.

Evidence-based medicine was (and still is) a challenge to some traditional medical practices. It has a distinctive rhetoric, with frequent use of words such as "rigorous" and "unbiased." The 1992 paper with which this chapter began makes repeated use of the phrase "methodological rigor." It also emphasizes the low value of knowledge based solely on authority. That is ranked at the lowest level of the hierarchy, when it is ranked at all. The general idea is that evidence-based medicine represents the admirable and democratic (because it is non-authoritarian) improvement of

---

[15] Some canonical definitions are less helpful, for example, that given in Sackett et al. (1996: 71): "Evidence-based medicine is the conscientious, explicit and judicious use of current best evidence in making decisions about the care of individual patients." This particular definition is probably used widely because it is brief, and because it has a rhetorical purpose—to address the frequent criticism of evidence-based medicine that it applies to populations, not individuals. David Eddy (2005) claims that it is the most commonly cited definition, and that is also my impression. Cohen et al. (2004), an excellent overview of the criticisms of evidence-based medicine, also prefers the Davidoff et al. (1995) definition.

standards of evidence in medicine, and that not following these methods takes medicine back to the dark ages of authority-based medicine and anecdotal (unsystematic, low quality) evidence. It is notable that from the perspective of evidence-based medicine, the methodology of consensus conferences—consensus of experts—looks more like knowledge based on authority than like the expert democratic process that it was also designed to be.

Practicing evidence-based medicine, as currently understood, requires access to (a set of guidelines based on) a systematic review (including meta-analyses where appropriate) for the intervention being contemplated. It requires an assessment that the current case is sufficiently similar to the trial situation(s) for extrapolation to be justified (this, especially, requires clinical judgment). It also requires consideration of patient preferences, and typically the tools of medical decision making are used for this (see Section 5.3). Finally, it requires an ongoing empirical mindset, with every case, if possible, providing data for future analysis.

The evidence hierarchy, comprised of "levels of evidence," is at the core of systematic reviews. Such hierarchies always rank evidence from high quality (by CONSORT standards) double-masked randomized controlled trials at the top or near the top of the hierarchy, but they also evaluate and make use of the full range of evidence. Evidence can include randomized controlled trials, observational trials (prospective and retrospective), cohort studies, case control studies, case series, individual case reports, and anecdotal evidence, usually in that order. Also sometimes (but not always) included in the evidence hierarchy, near the bottom, are expert opinion, background information, mechanistic reasoning, and pathophysiological rationale. (I think that calling these considerations "evidence" is misleading; more about that in Section 5.5) Evidence hierarchies are typically stated in a grid or a graphic (often a pyramid with the highest-quality evidence at the top and the lowest-quality evidence at the bottom). The GRADE evidence hierarchy has four levels: high, moderate, low, and very low. Rank in the evidence hierarchy is only one measure of the quality of a trial; it is also important to consider how well the trial lives up to the standards of its own place in the hierarchy, what the magnitude of the experimental effect is, and to what degree the conditions of the trial apply in the wider clinical context. For example, in the GRADE system of rating a high-quality observational trial can, overall, rank higher than a randomized controlled trial of lesser quality. (In more traditional systems of rating, this is not the case, and only randomized controlled trials can receive the highest ranking.) In 2002, the AHRQ reported 40 systems of rating in use, six of them within its own network of evidence-based practice centers (AHRQ 2002). Ross Upshur (2003) is just one of many researchers who have called for standardization. The GRADE Working Group, established in 2000, is attempting to reach consensus on one system of rating the quality and strength of the evidence. This is an ironic development, given that evidence-based medicine sees itself as replacing expert group consensus judgment.[16]

In his Harveian Oration, Michael Rawlins (chair of NICE in the UK) stated that "The notion that evidence can be reliably placed in hierarchies is illusory" (Rawlins 2008: 2152). This indicates his lack of confidence in the rigor of criteria for assessing levels of evidence. But even if we had clear criteria for assigning evidence to hierarchies, we would still have to make overall recommendations based on aggregated evidence. Such recommendations depend not only on the quality of the evidence but on its consistency, the magnitude of the effect, and the balance of benefits and risks for the intervention being considered. Recommendations themselves can be strong or weak (in the GRADE system; other systems have more nuances).

Evidence can sometimes be aggregated by the formal methods of meta-analysis, which combines the data from several similar high-quality trials to get an overall measure of the evidence. In a meta-analysis, it is standard to leave the data from lower-quality trials out of the calculation, although in principle it could be included and perhaps weighted less than the data from high-quality trials. Meta-analysis requires judgments about the similarity of trials for combination and the quality of evidence in each trial, as well as about the possibility of systematic bias in the evidence overall, for example due to publication bias and pharmaceutical company support (see Chapter 6). Meta-analysis is a formal technique, but not an algorithmic one. Different meta-analyses of the same data (using both the same and different systems of ranking) have on occasion produced different conclusions (Juni et al. 1999, Yank et al. 2007, Stegenga 2011). This has disturbed some

---

[16] Ironic, but also profound, because the foundations of evidence-based medicine are not (or not only) evidential. Instead, they may be, at least in part, pragmatic or arbitrary.

advocates of evidence-based medicine. For example, Steve Goodman thinks that conflicting meta-analyses of the data on screening mammography constitute a "crisis" for evidence-based medicine (Goodman 2002).

When meta-analysis is not an option (perhaps because there is evidence at several different levels on the hierarchy, or because trials differ too much for their data to be meaningfully combined), evidence needs to be aggregated in some different way. There are (mathematically speaking) infinite ways in which aggregation of evidence from different levels can be done, each way depending on the weights assigned to each particular level of evidence. Most aggregate assessments are not very fine grained, so different methods of aggregation may converge in their outcomes. GRADE, for example, makes either a weak recommendation or a strong recommendation whenever it finds that an intervention is effective. Such an aggregate assessment depends not only on the uncertainty about effectiveness, but also on assessments of the magnitudes of likely benefits and harms. There will be variability in these calculations, depending on both the preferences of patients and the considerations included by those assessing the evidence.

A final source of variability in systematic evidence reviews is due to the search tools used to identify the relevant literature. Different tools may be used, and the same tools used in different ways (e.g. with different search terms). Some searches will include non-English language publications and some will not. That said, there is a good deal of overlap in the results of searches using different methods. We are in a vastly different situation from that of Cochrane and Chalmers, who had to do their searches before the electronic age.

When evidence is strong and univocal, and assessment of benefits and harms is clear, the results of different meta-analyses and systematic reviews will also be clear and univocal. But such situations are not the intended material for these formal methods. Rather, we designed the methods to help us with the cases in which the direction and significance of the evidence is less obvious. Indeed, meta-analyses and systematic review may help with some less obvious cases, but in the least obvious cases (such as the case of screening mammography, which will be discussed in Chapter 9) different meta-analyses and different systematic reviews will reach different results and thus not reduce the uncertainty.

## 5.3 The Medical Decision Making Movement

Evidence-based medicine and medical decision making are partially overlapping areas of expertise that developed in different medical communities. Medical decision making started about a decade earlier than evidence-based medicine. To the extent that a clinician doing evidence-based medicine needs to calculate the likely possible harms and benefits of an intervention, she is dependent on the techniques of medical decision making. Medical decision making makes use of Bayesian probability theory and an investigation of harms, benefits, and individual variation in those harms and benefits. Shared decision making, in which the patient inputs their own preferences, can be used.

Medical decision making was a response to the literature of bias and error associated with Amos Tversky, Daniel Kahneman, and others. It has a similar ethos as evidence-based medicine, emphasizing quantitative rigor and precision, and the elimination of biases that get in the way of objective decision making. It has professional societies (e.g. the Society for Medical Decision Making, established 1978) and journals (e.g. *Medical Decision Making*, established 1981). It has not achieved the prominence of evidence-based medicine, for reasons that would be interesting to explore.[17]

The risk/benefit calculations can be made for individual patients, incorporating patient judgments of utility, or they can be made for populations, in the context of health care economics, using accepted standards for utilities. Medical decision making seeks to avoid common errors of judgment, such as overestimation or underestimation of risk, framing biases, and availability and salience biases.

While evidence-based medicine came from epidemiology and made use of developments in statistics, medical decision making came from internal medicine and health policy and made use of advances in the decision sciences, including social psychology. The movements share a commitment to the avoidance of bias, and an accompanying rhetoric of analytic rigor. Their techniques are quantitative, and require expertise in statistics and/or probability theory. Their techniques fit together: practicing evidence-based medicine requires the tools of

---

[17]  Arthur Elstein (2004) makes some intriguing remarks about this.

medical decision making and vice versa. Evidence-based medicine focuses on what we know, while medical decision making focuses on judgment. Evidence-based medicine weighs research evidence, while medical decision making weighs clinical evidence. Yet evidence-based medicine is much more prominent than medical decision making, and the techniques of medical decision making have sometimes been subsumed under it. Moreover, evidence-based medicine has often given short shrift to the importance of patient preferences and shared decision making.[18] Not surprisingly, this has been resented within the medical decision making community (Elstein 2004).

It may prove useful, at a future time, to take as close a look at the medical decision making movement as I have at the evidence-based medicine movement. There is certainly room for a good deal of historical and philosophical work on the two movements and their relations to each other.

## 5.4   A Common Criticism, and its Historical Antecedents

There are many criticisms of evidence-based medicine. Critics have protested, variously, that evidence-based medicine overlooks the role of clinical experience, expert judgment, intuition, the "art" of medicine, medical authority, patient variability, patient goals and values, local health care constraints, and the basic medical sciences, including the importance of theory and the understanding of causal relationships. This is a long and complex list of intertwined scientific, hermeneutic, political, and ethical considerations. In this chapter I will discuss just one of the criticisms of evidence-based medicine: the recurrent claim that evidence-based medicine ignores the basic sciences that guide both research and clinical practice (Ashcroft 2004, Bluhm 2005, Charlton & Miles 1998, Cohen et al. 2004, Harari 2001, La Caze 2011, Tonelli 1998). This is an especially important criticism because it leads to revealing the

---

[18]   Because of this, it is sometimes seen as paternalistic or authoritarian; indeed, despite its rejection of medical knowledge based on authority it seems comfortable with an authoritarian relationship between physician and patient. Evidence-based medicine is primarily driven by its epistemic goals of using the highest quality evidence, rather than by its political characteristics. The discussion of axes of variation in note 24, Chapter 2, also applies to the characteristics of evidence-based medicine, with democracy/authority as one of the axes of variation.

scientific characteristics of evidence-based medicine. Other criticisms will be discussed in Chapter 6 and Chapter 8.

Randomized controlled trials and other population-based studies are designed to discover whether or not health care interventions are effective. They are *not* designed to discover *how* health care interventions work (when they do work), or to come up with new ideas about mechanisms, new theories about disease processes, or new technologies for medical interventions. Health care interventions are judged effective when there is a correlation between the intervention and positive outcomes. Often it is not too much of a leap to infer that the intervention causes the positive outcome.[19] But the resulting knowledge is rather impoverished: it is knowledge of what works, without knowledge of *how* it works (or why it does not work), or how to make it work better. It is knowledge of effects without knowledge of underlying mechanisms.

There is a good reason for evidence-based medicine's skepticism about scientific theories. Scientific theories posit underlying mechanisms, but they often fail to make correct predictions about how those mechanisms work. There are countless examples of proposed interventions that make scientific sense and sometimes even work in vitro or in animal studies, but which turn out to be ineffective in humans. For example, hormone replacement therapy was expected to decrease the incidence of heart disease via the action of estrogens to reduce blood levels of harmful lipids, but actually increases mortality due to cardiac causes (this example is described at length in Howick 2011a; the appendix to chapter 10 in Howick 2011b contains many more examples). Other examples are bone marrow transplantation for metastatic breast cancer, early attempts at gene therapy, angiogenesis inhibitors for cancer, and vertebroplasty for osteoporotic spinal fractures.

The kind of medical knowledge produced by evidence-based medicine is what is often called "empiric medicine" by clinicians. This is knowledge of "what works" and "what does not work" in the treatment of symptoms and disease. This kind of knowledge actually has a venerable history in the Empiric school of medicine, starting in the third century BCE. The Empiric school developed in opposition to the Dogmatic school of Hippocratic medicine. The Dogmatic school aimed for a

---

[19] The precise conditions under which causation can be inferred from correlation are not of importance to the present discussion.

full understanding of the causes of disease (in those days, in terms of humors), and used this understanding to develop "rational" remedies (that is, remedies that were theoretically justified) whose outcomes were, unfortunately, not always carefully evaluated. The Empiric school was skeptical of humoral and other theories, or at least of their applicability to human illness in its full complexity, and limited themselves to therapies that were proven through clinical experience (which is not as rigorous as clinical trials, but is capable of detecting reproducibly strong evidence). Evidence-based medicine is of course much more sophisticated than Greek Empiric medicine at discerning what works and what does not work—its statistical, epidemiological approach is perhaps the most important difference—but it shares with Greek Empiric medicine a skepticism about drawing conclusions from our limited understanding of the mechanisms[20] of disease. (Evidence-based medicine also differs from Greek Empiric medicine in that it is practiced in the context of modern knowledge of disease classifications and diagnoses. Greek Empiric medicine also used the medical concepts of the day, and Empiric medicine is never completely free of theory.)

Bluhm and Borgerson (2011) describe evidence-based medicine as "empiricist" and contrast it with "rationalist" epistemologies of medicine. Note the slight difference in terminology—"empiricist" rather than "empiric"—which evokes a somewhat different philosophical background, that of the seventeenth and eighteenth century debate between empiricist philosophers (e.g. Locke, Berkeley, Hume) and rationalist philosophers (e.g. Descartes, Leibniz, Spinoza). I believe that the difference between "empiric" and "empiricist" is important here; "empiric" marks an eschewing of theory, while "empiricist" only marks the rejection of the a priori knowledge espoused by rationalists. Empiricist philosophers and scientists (Locke, Galileo, Newton, etc.) in the seventeenth and eighteenth century made extensive use of abstract theory. "Empiric" is a more precise characterization of the philosophical character of evidence-based medicine, connecting its methodology more with Greek medical traditions than with early modern thought.

---

[20] "Mechanisms" is intended broadly here to cover any kind of causal account of disease. (I am *neither* reading seventeenth-century mechanism *nor* the twenty-first-century Machamer, Darden, and Craver account of mechanistic process (Machamer et al. 2000) back into Greek thought!).

The contemporary term "empiric therapy," used to describe therapies that are prescribed when physicians do not have a complete diagnosis of an illness (or an understanding of the illness diagnosed) most likely also derives from the Greek Empiric school.[21] Empiric therapies address symptoms or broad categories of disease. For example, a broad spectrum antibiotic may be prescribed for an apparent bacterial infection (without doing any culturing or identification of the bacteria). Or, for example, chest percussion—a method developed for use in all kinds of bronchitis—was used in the treatment of cystic fibrosis in order to loosen chest congestion long before the basic mechanisms of cystic fibrosis were identified.

The philosopher Richard Ashcroft is unusual (among philosophers) but insightful in regarding it as an *advantage* that evidence-based medicine is "autonomous of the basic sciences" and "blind to mechanisms of explanation and causation (2004: 134). He regards it as an advantage because it means that evidence-based medicine does not have to worry that our basic theories may be incorrect. Ashcroft allies himself with Nancy Cartwright's realism about phenomenal laws and antirealism about deeper laws.[22]

Most commentators, however, see the eschewing of scientific theorizing as problematic (Bluhm 2005, Charlton & Miles 1998, Cohen et al. 2004, Harari 2001, La Caze 2011, Tonelli 1998, Howick 2011a, 2011b). Whatever one's views about scientific realism, evidence-based medicine typically depends upon a background of basic science research that develops the interventions and suggests the appropriate protocols. It is rare for an intervention without physiological rationale to be tested. An exception to this is when an intervention is accidentally discovered.[23] Moreover, "scientific judgment" enters into the design of appropriate randomized controlled trials (choice of control, eligibility requirements for participants, etc.) and into the assessment of the applicability of results to new cases (external validity). Such "scientific judgment" includes substantive knowledge of the basic mechanisms under study. Of course, many interventions with excellent physiological

---

[21]  I do not have any philological evidence for this, however.

[22]  Cartwright is, of course, a realist about underlying causal processes.

[23]  The discovery of lithium as a treatment for bipolar disorder is an example of this. It was discovered accidentally in the 1940s and we still do not know how it works.

rationales and good in vitro and in vivo performance fail when tested in human beings or fail when tested for external validity. That does not mean that there is anything wrong with using physiological reasoning (as far as it goes) or that we can use a more reliable method. It means that the physiological reasoning is inconclusive because there are often unknown confounders, additional mechanisms at play, and so on. The basic science work is highly fallible, but it is not dispensable. Even Nancy Cartwright (1989) would agree that we cannot replace physical theory with phenomenal laws alone.

It may be illuminating at this point to recall Claude Bernard's 1865 *An introduction to the study of experimental medicine* (Bernard 1957). Claude Bernard was one of the founders of scientific medicine, and he wrote at some length about methodology. He wanted to go *beyond* what we call "empiric medicine," in order to find out how physiological processes work. He called physiological processes "the hidden springs of living machines," and claimed that living creatures have no special "vital" substance; rather, they obey the same physical laws as inanimate matter. He saw experimentation, especially laboratory work, as the best method of discovering the mechanisms by which physiological processes work. And he saw the discovery of mechanisms as a necessary preliminary to inventing effective treatments. Thus evidence-based medicine—seen by its advocates as the most objective and highest-quality methodology for clinical research—does not live up to the standards of Claude Bernard's scientific medicine. (This is a nice example of the different meanings and uses of the term "scientific.")

How can evidence-based medicine respond to the criticism that it devalues knowledge of mechanisms? One option is to argue that evidence-based medicine, properly understood, regards (or should regard) pathophysiological (and/or mechanistic) understanding as *a kind of evidence* with an appropriate place(s) on the evidence hierarchy. This is the option taken by Jeremy Howick (2011a, 2011b), Holly Andersen (2012), and Brendan Clarke et al. (2013). I will take a look at this view in Section 5.5, focusing on Jeremy Howick's arguments for the sake of clarity and conciseness. Although there are differences between Howick (2011b), Andersen (2012), and Clarke et al. (2013), they all share the assumption that mechanistic reasoning should be regarded as mechanistic evidence, moreover, evidence that has a place in the hierarchy. This is an assumption that I challenge.

Another option is to acknowledge the importance of mechanistic (and/or pathophysiological) reasoning and to conclude that evidence-based medicine is not a *complete* epistemology of medicine, in that it cannot be done without other methods such as mechanistic reasoning. This is the option that I will eventually take.

## 5.5  Treating Mechanistic Reasoning as Mechanistic Evidence

Howick tries to bring mechanistic reasoning under the framework of evidence-based medicine by treating it as a kind of evidence (2011a, 2011b). Most evidence hierarchies place "pathophysiological rationale" (which for our purposes can be equated with mechanistic reasoning) at the bottom, as the lowest level of evidence—or leave it off the hierarchy altogether. This puts mechanistic reasoning at approximately the same level as anecdotal evidence and expert consensus, which receive similar treatment and arguably are not evidence at all. Howick thinks that this does not do justice to the importance of mechanistic reasoning, and to its strength in some instances. In fact, he thinks that "mechanistic reasoning" provides "mechanistic evidence," which he regards as a kind of evidence, just as observational trials are a kind of evidence. Howick argues that "mechanistic evidence" does not have a fixed place in the evidence hierarchy; sometimes it is high quality and sometimes it is low quality. This means that it does not always belong at the lowest level of evidence, along with expert consensus and anecdotal evidence. He would prefer to rank it higher in the evidence hierarchy, especially when it is "high-quality" mechanistic reasoning. Howick defines "high quality mechanistic reasoning" as "valid and based on 'complete' mechanisms" (2011a, 2011b). He gives an example of such high-quality mechanistic reasoning: the proposed use of radiotherapy to shrink goiters and thereby improve respiratory function at a time when it was already known that radiotherapy shrinks goiters without harmful side effects. So Howick thinks that when we have "complete" mechanistic knowledge we can have high-quality mechanistic reasoning and thereby high-quality mechanistic evidence.

I have a couple of difficulties with this proposal for including mechanistic reasoning in the evidence hierarchy. The first is that the example of high-quality mechanistic reasoning, as stated, is logical rather than mechanistic reasoning. We are given that radiotherapy shrinks goiters, that large goiters impair respiratory function because they obstruct the airway, and that radiotherapy has no paradoxical responses or harmful side effects. The reasoning here could be presented as a deductive argument with the conclusion that radiotherapy will improve respiratory function in people with large goiters that obstruct the airway:

Premises

(1)  For all x, if x is a goiter then x can be shrunk by radiotherapy
(2)  Large goiters impair respiratory function
(3)  Small goiters do not impair respiratory function
(4)  Radiotherapy for goiters is safe

Consider a large goiter G that impairs respiratory function. Following premise (1) and premise (4) it can be safely shrunk by radiotherapy until it is a small goiter, and, following premise (3) we expect that:

Conclusion

Respiratory function will be improved if large goiters are treated by radiotherapy

So this is not really an example of mechanistic reasoning; it is an example of deductive logic. Perhaps that is why Howick uses the word "valid" in stating the properties of high-quality mechanistic reasoning! The truth of the conclusion, of course, depends on the truth of the premises, which rest on inductive inference, so we do *not* have, overall, deductive certainty that radiotherapy for large goiters will improve respiratory function. The need for a clinical trial of the effects of radiotherapy on goiter to improve respiratory function should depend on the strength of our knowledge of the premises—not only on the strength of our reasoning from premises to conclusion.

So Howick owes us an account of mechanistic (rather than logical) reasoning, one that can justify assessments that some cases of mechanistic reasoning are stronger than others. He suggests that we do so in terms of the *completeness* of our knowledge of relevant mechanisms, which means understanding their complexity and their stochastic nature (Howick 2011b). That is a plausible suggestion. It is difficult, however, to assess our knowledge of the completeness of relevant mechanisms. We certainly have a historical tendency to expect greater simplicity than there is (as we did in the early years of the Human Genome Project, when one gene to one disease correlations were widely expected). It may also be that mechanistic reasoning is never very strong, or that there are no real time (not hindsight) objective assessments of its strength.[24]

The second difficulty with the proposal to include mechanistic reasoning in the evidence hierarchy stems from the fact that the term "mechanistic reasoning" is not equivalent to the term "mechanistic evidence." Howick uses the terms interchangeably, and I think that this leads to confusions. To be sure, we can have evidence for mechanisms, but that is evidence that the mechanisms operate, not evidence that a particular proposed intervention (which depends on more than the hypothesized mechanisms, even if those mechanisms exist) will work. We could have strong evidence that the mechanisms operate, yet no evidence (or the weakest of evidence) that a particular proposed therapy will have the desired effect.[25] Howick's example of the initial expectation that hormone replacement therapy would reduce cardiac mortality is a good example of this. We had strong evidence of hormonal effects on blood lipids, but weak (perhaps even no) evidence that this clinical intervention would reduce cardiac mortality, because we did not have full knowledge of the relevant complexity of mechanisms. Moreover, *we did not know that we did not know* the relevant complexity of mechanisms, and thus exaggerated the little evidence there was for expecting hormone replacement therapy to reduce cardiac mortality. As the evidence came in, it became clear that hormone replacement therapy is associated with an increase in cardiac mortality.

In his review of Howick's book, Alexander Bird phrases the issues more carefully: he describes Howick's project as an inquiry into "the *evidential* role of mechanistic (pathophysiological) reasoning" (Bird 2011: 645) (author's italics). It is implied (and I agree) that mechanistic reasoning can play roles other than an evidential role. In fact, I will argue that mechanistic reasoning provides weak evidence at best, but has

---

[24] See also Bluhm (2011), which makes similar points about Howick's need to say more about how to modify the standard evidence hierarchy to include "high-quality" mechanistic reasoning.

[25] Whether we say that there is no evidence or the weakest of evidence depends on our threshold for counting something as evidence.

important non-evidential roles. In particular, I will argue, mechanistic reasoning is a tool of discovery.

What is at stake with whether or not we value mechanistic reasoning for its evidential role or for some other role? I think what is at stake is whether or not evidence-based medicine is a complete epistemology of medicine. If mechanistic reasoning does not count as good evidence, then, since mechanistic reasoning is clearly necessary and important, evidence-based medicine is not a complete epistemology of medicine. Howick is a champion of evidence-based medicine, so it is not surprising that he tries to see it as a complete epistemology. I will propose a different way of understanding evidence-based medicine, one in which mechanistic reasoning has an important but non-evidential role in a more complete epistemology of medicine.

## 5.6　The Place of Evidence-Based Medicine in a Complete Epistemology of Medicine

The early practitioners of evidence-based medicine thought that they had a new kind of methodological rigor, and that older methods such as clinical intuition, expert consensus, and pathophysiological rationale were much more fallible. They may be correct about this.[26] But we do not have the option to use evidence-based medicine everywhere and thereby get more reliable and less fallible science. Evidence-based medicine cannot get going unless a therapy is proposed. How do we come up with new therapies? Typically, we use pathophysiological reasoning and preliminary trial and error experimentation. We go through brainstorming and modeling, and then in vitro trials, animal in vivo trials, Phase I and Phase II clinical trials to determine safety and efficacy, before getting to the Phase III trials that provide the evidence of effectiveness for evidence-based medicine conclusions. Often, proposed therapies fail before Phase III. We may wish that we had more reliable methods of discovery, but evidence-based medicine does not help us with them. (In fact, I will argue in Chapter 7 that translational medicine is an effort to help us with discovery.)

---

[26] But see Chapter 6. Evidence-based medicine is more fallible than expected.

I think of mechanistic (pathophysiological) reasoning and early experimentation as part of what logical empiricists used to call "context of discovery." Hans Reichenbach(1938) introduced the term, and the associated distinction between "context of discovery" and "context of justification." The idea is that the two contexts call for different methodologies. The "context of discovery" allows any creative methods but the "context of justification" is the place for rigor in evaluating the creative ideas developed in the context of discovery. It is a romantic view of science in which scientific advance depends on both unconstrained imagination and on rigorous logic. Since the time of the logical empiricists, we have come to appreciate that the context of justification is not so rigorous, and that the context of discovery is not so unconstrained.[27] One of the tools of discovery is thinking about mechanisms; another is early experimentation. In the context of discovery, possible interventions (for diagnosis or treatment) are proposed. Evidence-based medicine can be fruitfully understood as the "context of justification" in which evidence for an intervention is gathered and evaluated.

So evidence-based medicine should not discount mechanistic reasoning, unless it wants to bite the hand that feeds it! In the more recent evidence-based medicine literature, I see some recognition of this. Donald Berwick's "Broadening the view of evidence-based medicine" (2005) is particularly insightful. (Berwick is the founder of the Institute for Healthcare Improvement, which is the leading organization for quality improvement in health care.) The paper claims that "we have overshot the mark" with evidence-based medicine and created an "intellectual hegemony" that excludes important research methods from recognition (Berwick 2005: 315). Berwick calls the overlooked methods "pragmatic science" and sees them as crucial for scientific discovery. "Pragmatic science" includes methods such as taking advantage of local knowledge, using open-ended measures of assessment, and "using small samples and short experimental cycles to learn quickly." Evidence-based medicine has not formalized all the tools that we need to do science, only some of them. Berwick recommends that we not only encourage, but

---

[27] See Lindley Darden (2006) and Nancy Nersessian (2008). In Solomon (2009) I argue that the logical empiricists' account of the "context of discovery" too weakly constrains thought, and that creativity needs more structure.

*publish* results in "pragmatic science." This theme will be taken up again in Chapter 7.

## 5.7   Case Study: Evidence-Based Medicine and Mechanistic Reasoning in the Case of Cystic Fibrosis

The purpose of this section is to present the case of recent research on cystic fibrosis as an extended example of how mechanistic reasoning is sometimes successful and sometimes not, at all points on the causal chain (or network) from "root causes" to production of symptoms. Note that despite its low reliability (as the epistemologist Alvin Goldman might describe the situation), mechanistic reasoning is indispensable to early-stage research. Early-stage research is not quite the same as the logical empiricists' "context of discovery," which focused on unconstrained theoretical imagination (sometimes done in an armchair while dozing in front of the fireplace); early-stage research includes reasoning constrained by mechanistic/causal knowledge and the results of trial and error experimentation.

Cystic fibrosis is an autosomal recessive genetic disorder that occurs primarily but not exclusively in people of European descent. It is one of the most common life-shortening genetic diseases (1 in 25 people of European descent are cystic fibrosis carriers; 30,000 people with cystic fibrosis live in the US). It was first identified in 1938 by the American physician Dorothy Andersen, on the basis of a cluster of clinical symptoms affecting the lungs, pancreas, liver, and other organs. Lung inflammation and frequent infections are the most serious problems, apparently caused by the buildup of thick mucus. Life expectancy has greatly increased, from a few months in 1950 to 37 years in 2008. The increase in life expectancy has been due to the gradual accrual of treatments addressing symptoms of cystic fibrosis, starting with antibiotics, chest percussion, and supplementary digestive enzymes, and continuing more recently with bronchodilators, saline mist, Pulmozyme, and, since 2012, genomic and proteomic treatments. The gene responsible for cystic fibrosis was identified on the long arm of chromosome 7 in 1989 by a team led by Francis Collins. It is called the CFTR gene—cystic fibrosis transmembrane conductance regulator—which codes for a protein regulating

chloride transport across cell membranes. The most common mutation is deltaF508 (two-thirds of cases), but there are over 1,500 known
mutations of the same gene.

Treatments for cystic fibrosis are complex and time consuming, taking several hours per day as well as regular professional monitoring and
adjustment. By the 1970s it became evident that cystic fibrosis patients do
best when taken care of by multidisciplinary teams (physicians, nurses,
social workers, physical therapists, etc.) at specialized cystic fibrosis care
centers, and this has become standard in the developed world. In the
United States, the Cystic Fibrosis Foundation plays an important role in
coordinating both research and care. Similar organizations operate in
other countries and Cystic Fibrosis Worldwide operates at an international level. The well-known surgeon/writer Atul Gawande praises these
institutions, writing that "cystic fibrosis care works the way we want all
of medicine to work" (2004). He is impressed with the systematic and
evidence-based delivery of health care. He also reports that in the best
treatment centers—those with aggressive adherence to evidence-based
guidelines—life expectancy is 47, ten years longer than average.

The staples of cystic fibrosis treatment are not particularly high-tech,
nor specific to cystic fibrosis treatment. They were developed for
other uses and qualify as "empiric therapies" in the sense discussed in
Section 5.4. Chest percussion several times daily to loosen mucus, oral
and inhaled prophylactic antibiotics, bronchodilators, hypertonic saline
mist, supplementary digestive enzymes, ibuprofen, Pulmozyme, and
aerobic exercise are recommended even for those who are asymptomatic.
For cystic fibrosis, it turns out, it is better to prevent symptoms than to
treat them when they occur. In advanced disease, additional intravenous
antibiotics, inhaled oxygen, ventilators, and even lung and liver transplantation are standards of care.

How were these therapies discovered? With the possible exception
of Pulmozyme,[28] they were not created specifically for cystic fibrosis, but for other conditions with similar symptoms. They address the
symptoms—the distal end(s) of the causal chain(s) starting with the
defective CFTR gene. They do not address the proximate end of the

---

[28] Pulmozyme is a non-specific therapy because it works by breaking down DNA. It is
also in use for breaking down the mucus in "glue ear."

causal chain: the defective gene and its products. Thus they are "empiric therapies" (sometimes also called "symptomatic therapies") in the sense discussed in Section 5.4. Each proposed intervention was tested, and clinical trials have improved in quality over time (as they have in the rest of medicine). The proposed empiric therapies were evaluated according to the standards of evidence-based medicine. Some similarly proposed interventions turned out not to be effective: for example, sulfa drugs for infection, high humidity (mist) tents for loosening mucus, and corticosteroids for reducing inflammation. There was no way to figure out in advance that other antibiotics would do better than sulfa drugs, hypertonic saline mist better than water mist, and ibuprofen better than corticosteroids. Sometimes we do not even understand the reasons for our successes and failures in hindsight. For example, no one understands yet why hypertonic saline mist works better than water mist, or why ibuprofen works better than corticosteroids. Our current knowledge about cystic fibrosis treatment is the product of more than 50 years of clinical trials. Our knowledge of cystic fibrosis therapies is evidence-based, in the precise sense intended by "evidence-based medicine." Of course, not all trials were perfectly randomized double-masked controlled trials, especially in the early years, but evidence-based medicine (properly understood) includes the observational and other kinds of trial that provided knowledge about effective interventions. Much recent research on empiric (symptomatic) therapies for cystic fibrosis has been carried out in double-masked randomized controlled trials producing the highest quality of evidence. The results of these clinical trials have been combined in systematic reviews such as a comprehensive pulmonary therapy review produced by the Cystic Fibrosis Foundation (Flume et al. 2007) and many recent Cochrane reviews of particular therapies. It is the implementation of this knowledge of empiric therapies in system-based care that Atul Gawande is commending.

At the same time as these improvements in clinical care for cystic fibrosis were discovered, knowledge of the genetic mechanisms producing the disease advanced in more basic research. After the identification of the cystic fibrosis gene on the long arm of chromosome 7 in 1989, there was widespread optimism that gene therapy replacing the defective CFTR gene with a normal CFTR gene was around the corner, and that cystic fibrosis would disappear before the end of the 1990s (Lindee & Mueller 2011). The goal was to correct the "root cause" of cystic fibrosis (i.e. the defective gene) by correcting a basic mechanism, the gene itself. There was success in both mouse models and in vitro models of gene therapy for cystic fibrosis in the early 1990s. Cystic fibrosis was one of the first genetic diseases for which gene therapies were designed, and early human trials used an inhaled adenovirus to try to transfer healthy CFTR genes to lung cells. None of these interventions made it to Phase III clinical trials: there were immediate problems with both safety and efficacy. The widespread confidence that a cure for cystic fibrosis was imminent faded by the late 1990s (Lindee & Mueller 2011). Researchers moved on to other genetic conditions, such as ornithine transcarbamylase (OTC) deficiency. The ornithine transcarbamylase trial at the University of Pennsylvania ended in 1999 with the death of an experimental subject, Jesse Gelsinger, and since then gene therapy research has been proceeding with more modesty and greater caution. We do not know how far away a solution to the technical difficulties of gene therapy is, or even if there is a solution.

Identification of the genetic mechanisms of cystic fibrosis has been applied more successfully in the development of carrier, prenatal, and newborn genetic testing. However, the result of this work has been to create *more* uncertainty about prognosis, treatment, and even definition of disease, since it turns out that clinical disease depends on both the type of mutation of the CFTR gene (there are more than 1,500 known types of mutation) and on other, mostly unknown, factors such as modifier genes. It turns out that there are individuals with two defective genes who do not suffer clinical disease at all, as well as individuals with one defective gene who have clinical manifestations of cystic fibrosis. It is not possible to predict the severity of the disease from the type of mutation (there is some correlation, but clearly other factors play a role), and there is considerable phenotypic variability (Lindee & Mueller 2011).

Lindley Darden said a few years ago, with some exasperation, that the genotype–phenotype relations in cystic fibrosis are "more complex than anyone studying a disease produced by a single gene defect had reason to expect" (Darden 2010). In fact, evidence of complexity was already there in the late 1980s, by which time it was known that cystic fibrosis exhibits considerable genetic and phenotypic variation, with poor correlations of genotype to phenotype. Even simple Mendelian inheritance does not mean simple molecular mechanisms. The tendency to oversimplify genetics has been with us since the birth of Mendelism, and perhaps reached its peak during the 1990s with the excitement surrounding the Human Genome Project and its anticipated commercial applications.

What is a mechanism? A standard account that is helpful to mention here is the Machamer, Darden, and Craver (MDC) account: "Mechanisms are entities and activities organized such that they are productive of regular changes from start or set up conditions to finish or termination conditions" (Machamer et al. 2000: 3). At a more general level, mechanistic theories are a common kind of scientific theory. Mechanistic explanations have a *narrative* form in that they are often linear, sequential, causal, and extended in time. They are typically deterministic narratives that, in the case of cystic fibrosis, have fuelled hope by highlighting apparent opportunities for interrupting the regular chain of events with causal interventions. This is the case for both the distal end of the causal chain, already discussed, and for the proximal end—the defective gene and the biochemistry of the CFTR protein. Although our knowledge of some basic mechanisms in cystic fibrosis has grown, we do not understand these mechanisms fully. For example, we do not know whether the excess mucus is produced by an inflammatory response or by chloride transport problems or both. Full mechanistic understanding of cystic fibrosis would of course include understanding the functional abnormalities in each of the more than 1,500 mutations that can cause the disease, as well as the role of modifier genes.

Despite the last 25 years of increasing understanding of the "root" mechanisms of cystic fibrosis, the gains in life expectancy have come from addressing the distal end of the chain of causes, and have not depended on knowledge of molecular mechanisms. They have depended on knowledge of macroscopic mechanisms behind symptoms. Evidence-based knowledge about cystic fibrosis care is, overwhelmingly, knowledge about non-specific ("empiric") and mostly low-tech interventions that increase lifespan from infancy well into middle age. So our most *precise* knowledge is about our *crudest* interventions. This epistemic irony is perhaps only temporary. Although so far gene therapy for cystic fibrosis has failed, there are other proposed interventions targeting basic mechanisms ("genomic" and "proteomic" therapies rather than "gene" therapies) that have shown promise in the last few years.

The first proposed intervention is the use of drugs designed to correct the mutated CFTR protein so that it can restore chloride transport to normal. With over 1,500 ways of making mistakes in the cystic CFTR protein, it is likely to be a lengthy project to complete. However, there is a promising start with the development of "chaperone" drugs such as Kalydeco (formerly VX770, from Vertex Pharmaceuticals), which is designed to repair the proteins produced by the G551D mutations (about 4 percent of cystic fibrosis patients have the G551D mutation). Kalydeco was FDA approved in January 2012. Recently a trial combining Kalydeco with VX809 showed effectiveness and safety for cystic fibrosis patients with the most common mutation, deltaF508 (about 66 percent of cystic fibrosis patients have at least one deltaF508 mutation). Currently a Phase II clinical trial is using Kalydeco together with another drug (VX661), also for the delta508 mutation.

Ataluren (PTC124, from PTC Pharmaceuticals) aims to be a general genomic fix for "nonsense mutations" by making ribosomes less sensitive to nonsense codons. It has been tested for Duchenne muscular dystrophy as well as cystic fibrosis. Although it has been tested in clinical trials, the results have been mixed and its effectiveness is uncertain.

Another idea for a cystic fibrosis therapy targeting basic mechanisms is to create or enhance an alternative chloride transport channel that does not depend on the CFTR protein. The drug Denufosol was designed to do just this, and did well in early Phase I and II clinical trials. If the drug had worked, it would probably have been appropriate for all cystic fibrosis patients, no matter which underlying mutation of the CFTR gene they have. Unfortunately, Inspire Pharmaceuticals announced in January 2011 that Denufosol did not show efficacy in Phase III clinical trials.

Even with apparently good mechanistic understanding, it is not possible to know in advance which proposed interventions will work. In this sense, there is no difference between intervening at the proximal or the distal end of the causal chain. We saw already that several reasonable suggestions about interventions at the distal end also did not work (sulfa drugs, mist tents, corticosteroids). A general problem with mechanistic accounts is that they are typically incomplete, although they often give an *illusion* of a complete, often linear, narrative. Incompleteness is the consequence of there being mechanisms underlying mechanisms, mechanisms inserted into mechanisms, background mechanisms that can fill out the mechanistic story, and mechanisms that can hijack regular mechanisms. That is, there is complex interaction of multiple mechanisms in a chaotic and multidimensional system. There are possible hidden mechanisms everywhere in mechanistic stories, despite an easy impression of narrative or causal completeness. Since we do not have a theory of everything, it is not possible to know in advance whether or not a particular mechanistic intervention will have the intended result. Mechanistic reasoning produces suggestions, which then have to be tested.

## 5.8 Conclusions

Evidence-based medicine is a contemporary version of Greek and traditional empiric medicine in putting the focus on finding therapies that work rather than on theories explaining disease. It is far more sophisticated than Greek and traditional empiric medicine at collecting and assessing the evidence for the effectiveness of a medical intervention. However, it is not a complete epistemology of medicine. Evidence-based medicine is often criticized for leaving out pathophysiological, mechanistic, and basic science reasoning. In my view, the solution is not to add these in to evidence-based medicine and try to find their place in the evidence hierarchy but, rather, to acknowledge that assessing the quality and strength of the evidence is not all there is to epistemology of medicine (or science).

The case study of the treatment of cystic fibrosis is a typical story of failures and successes in medicine. Mechanistic reasoning is used at all places on the causal chain of pathology to try to intervene with the disease process. By a trial and error process, interventions were tried and tested. The earliest successes came from interventions at the distal end of the causal chain, but in the last couple of years we have seen some initial successes based on intervention at the level of basic mechanisms. Mechanistic reasoning has been indispensable in the process of discovery despite the fact that its reliability is low because most interventions do not work. Mechanistic reasoning (or "mechanistic evidence") does not play a role in the process of evaluating the effectiveness of new interventions.

# 6

# The Fallibility of Evidence-Based Medicine

## 6.1 Introduction

The goal of this chapter is to explore the ways in which evidence-based medicine, while in several ways methodologically superior to more informal evidence collection and appraisal, is more fallible than generally acknowledged or expected.[1] It is important to get a feel for the characteristics of this fallibility because in some contexts, such as health care policy, evidence-based medicine plays a hegemonic role that may not always be justified. In particular, I challenge the implicit assumption that a hierarchy of *absence of biases* (with the evidence having the fewest potential kinds of bias at the top) amounts to a hierarchy of *reliability of evidence*. This is an assumption because the potential and even the presence of bias indicate the possibility, not the actuality, of error. Moreover, the number of *types* of bias is not equal to the overall *magnitude* of bias. This chapter also makes modest suggestions for improvement of the reliability of evidence-based medicine, not so that it can occupy a hegemonic role among methods, but so that it can do better.

Four different kinds of critique have cast doubt on the reliability of evidence-based medicine. First, philosophers and other theoreticians have argued that the double-masked randomized controlled trial, while certainly freer from bias than other trial designs, is not bias free. So even this "gold standard of evidence" is not sufficient for avoiding bias in research. Second, philosophers and epidemiologists

[1] Some portions of this chapter are adapted from Solomon (2011b).

have pointed out the ways in which practicing evidence-based medicine (which involves applying the results of randomized controlled trials and other trials) is dependent on more general scientific knowledge and judgments and is therefore vulnerable to any deficits in that knowledge and those judgments. Third, some epidemiologists have measured the reliability of randomized controlled trials empirically and found that they are less reliable than expected, suggesting that some biases have been overlooked. Finally, different evidence hierarchies and different standards for aggregation of evidence in meta-analysis sometimes yield different assessments of the weight of a body of evidence, suggesting that some assessments are more arbitrary than they appear (or perhaps incorrect). I will explore these critiques in Sections 6.2 to 6.6 (taking two sections, 6.3 and 6.4, to discuss the second kind of critique).

## 6.2 Internal Validity: Procedural Biases

The double-masked randomized controlled trial is designed to avoid many methodological biases, including selection (allocation) bias, confusion of specific drug effects with placebo effects, availability and salience biases, and confirmation biases. Many of the requirements of a well-designed randomized controlled trial such as adequate size of the trial population, the importance of a control arm(s), the need for masking, and the requirement for prior specification of the endpoint(s) have been learned through past experiences of false and misleading outcomes in clinical trials. Current standards of trial methodology are the result of almost a century of work in statistical methods.

John Worrall (2007b) and Brendan Clarke et al. (2013) have argued that the double-masked randomized controlled trial is not sufficient to guarantee results that are free of selection bias, even though it is designed to do just that. Worrall (2007a), Jeremy Howick (2011b), and many others (e.g. Smith & Pell 2003, Glasziou et al. 2007, Schunemann et al. 2008) have also argued that a double-masked randomized controlled trial is not necessary for producing strong evidence.[2] This section will take a look at these arguments in turn.

---

[2] Others (such as Rawlins 2008) have made the same point, but Worrall and Howick are the philosophers of science who are best known for this line of thought.

### 6.2.1 The Double-Masked Randomized Controlled Trial Is not Sufficient for Producing Strong Evidence

Worrall argues that randomized controlled trials do not eliminate all possibilities for bias, and in particular are vulnerable to selection bias despite the use of randomization. Moreover, he suggests that this potential for bias is, in practice, ineliminable. If he is correct, then the randomized controlled trial is not sufficient for the elimination of significant bias. The core of Worrall's argument (Worrall 2007a, 2007b, 2002) is that randomization can reduce but does not eliminate selection bias. The problem is that randomization can control only for most, but not for all, possible confounding factors. (A confounding factor is a factor that disproportionately affects either the experimental or the control group, thus leading to selection bias.) When there are indefinitely many factors,[3] both known and unknown, which may lead to bias, it is likely that any one randomization will not randomize with respect to *all* these factors. Under these circumstances, Worrall concludes, it is likely that any particular clinical trial will have at least one kind of bias that makes the experimental group relevantly different from the control group—just by accident. The only way to avoid this is to re-randomize and do another clinical trial, which is likely, again by chance, to eliminate the first trial's confounding bias but introduce another. Worrall concludes that the randomized controlled trial does not yield reliable results unless it is repeated time and again, re-randomizing each time, and the results are aggregated and analyzed overall. The number of repeats required is far larger (it is indefinitely large) than the usual number of replications of a study (usually, at best, a number in the low single digits). This much replication is, in practice, impossible.

In context, Worrall is less worried about the reliability of randomized controlled trials than he is about the assumption that they are much *more* reliable—in a different epistemic class—than well-designed observational ("historically controlled") studies, for example, in which there is no randomization. He is arguing that the randomized controlled trial should be taken off its pedestal and that *all* trials can have inadvertent bias due to differences between the control and the experimental group.

---

[3] Worrall refers to Lindley (1982) for the argument that there are indefinitely many factors.

While this is so, I think that Worrall is making more of a logical point than a practical one. Randomization is a powerful technique that usually controls for confounding factors. It is unlikely to control for confounding factors only in the event that there are *many unrelated* population variables that influence outcome, because only in that complex case is one of those variables likely to be accidentally unbalanced by the randomization. Worrall considers only the abstract possibility of multiple unknown variables; he does not consider the likely relationship (correlation) of those variables with one another and he does not give us reason or even specific cases to argue that, *in practice*, randomization generally (rather than rarely) leaves some *causally relevant* population variables accidentally selected for and thereby able to bias the outcome.[4] He gives no actual examples in which randomization failed to control for selection bias. In addition, the common requirement for at least one successful replication[5] can add evidence that any confounder inadvertently introduced is not causally responsible for the outcome.

Worrall's more general point—that the double-masked randomized controlled trial is not guaranteed to be "bias free"—is, however, well taken. No one has designed a "bias free" trial methodology and indeed such a methodology is probably unachievable. Even if trials are free of selection bias, they are vulnerable to many other sorts of biases, such as confirmation bias, funding bias, self-selection bias,[6] data-fishing bias, and inappropriate entry criteria. The double-masked randomized controlled trial has proved, through experience, to be a useful trial methodology for certain sorts of therapeutic interventions, especially pharmaceutical interventions with modest, variable, and subjectively measured outcomes.

Grossman and Mackenzie (2005) have argued that appreciation of the wide range of possible kinds of bias shows that the usefulness of the evidence hierarchy is limited. The evidence hierarchy says only that "all other things being equal" randomized controlled trials have less bias

---

[4] Worrall might respond that he is only criticizing those methodologists who make abstract and general claims about freedom from "all possible" biases. This is fine, so long as no conclusions are drawn for randomized controlled trials in practice.

[5] This is not a universal requirement but it is a reasonable and a practical one and is used by, for example, the FDA in the United States.

[6] By this is meant the decision to participate in a randomized controlled trial (not the decision about which arm to take).

than observational trials and other sorts of evidence. But all other things are *never* equal; for example, observational trials are less affected than randomized controlled trials by self-selection bias, and better positioned to detect adverse effects of treatment. Moreover, most trials have methodological flaws of one kind or another, either in design or in performance. Most hierarchies of evidence still rank a flawed randomized controlled trial above a high-quality observational trial, and this practice is questionable.[7] The hierarchy of evidence is not a hierarchy of reliability or trial validity; it is, at best, a hierarchy of freedom from some sorts of potential bias.

### 6.2.2  *The Double-Masked Randomized Controlled Trial Is not Necessary for Producing Strong Evidence*

It has frequently been pointed out (e.g. Smith & Pell 2003, Glasziou et al. 2007, Schunemann et al. 2008) that a randomized controlled trial is not needed for some medical interventions: those for which we already have strong-enough evidence. These are typically cases in which treatments have consistent and large effects. Examples given are ether for anesthesia, insulin for diabetes, blood transfusion for severe hemorrhagic shock, tracheostomy for tracheal obstruction, defibrillation for ventricular fibrillation, and suturing for repairing large wounds. It should be noted that in these cases lack of treatment often has dire effects and treatment has acceptable side effects; we make such determinations without randomized controlled trials but not a priori.[8] "Strong-enough evidence" is the ultimate goal in medical research; "levels of evidence" are of relevance to this assessment only when we do not know whether or not we have strong-enough evidence, and when we have confidence that levels of evidence grade strength of evidence.

Howick (2011b, 2008) argues against the common presumption that the double masking is the ideal for all trials of medical therapy. He shows that the requirement for masking is often either unnecessary or impossible to fulfill. Masking is indicated only when effect size is small, variable,

---

[7] GRADE has a scoring system in which it is possible for an observational trial to rank higher than a randomized controlled trial. It remains to be seen whether this system will be widely adopted and show inter-rater reliability.

[8] Even the much-cited parachute example (Smith & Pell 2003) rests on our experience that parachute use is not associated with unacceptable consequences (such as killing others with rope entanglements or landing injuries).

and subjectively measured—so small and variable that any apparent effects could be due to bias in outcome measures (leading to confirmation bias) or simply to the placebo effect.[9] When effect size is large and/or consistent, masking is impossible because the effects of the drug unmask the assignment. This is also true when side effects are large and/or consistent. And when outcomes are measured objectively (in terms of life expectancy, say, rather than reported level of pain) the confirmation biases that masking is designed to avoid are unlikely, making masking unnecessary for avoiding confirmation bias.

Howick is correct that masking is feasible and important only for detecting small effects with variable and subjectively measured outcomes and side effects. However, this is in fact the case for many recent advances in medicine, so that masking is frequently desirable and achievable. We can even do "sham surgery," which involves minimal intervention, in order to improve the quality of clinical trials to determine the effectiveness of some surgeries.[10] It is not often that we have new interventions with the dramatic success of insulin for diabetes or vascular surgery for ruptured aortic aneurism. Howick is right to point out that the methodology of the double-masked randomized controlled trial is suited for some interventions and not suited for others, but the methodology is, in fact, suited for many if not most of the health care interventions currently in development. In addition, sometimes masking is desirable but practically impossible, such as when interventions require a strict diet,[11] when treatments have characteristic side effects, or when the process of recovery from surgery reveals the surgical intervention used.

Howick (2008) also cites studies (e.g. Fergusson et al. 2004) showing that masking is often unsuccessful because trial subjects often (more often than would be predicted by chance) correctly guess which

[9] The placebo effect is generally small. (If only it were larger!) Masking is important when the magnitude of the therapeutic intervention is similar to the magnitude of the placebo effect.

[10] Surgery is often impossible to mask because the need for recovery after real surgery can unmask the intervention. Cases of successful masking in surgery are minimally invasive procedures such as arthroscopic knee surgery for osteoarthritis (Kirkley et al. 2008) and vertebroplasty for osteoporotic spinal fractures (Kallmes et al. 2009, Buchbinder et al. 2009).

[11] This was a problem early on with the use of double-masked randomized controlled trials in medicine, as Harry Marks (1997) has documented for studies of diet and heart disease.

treatment arm they have been assigned to—not because of the success of the intervention arm, but for other reasons such as the characteristic side effects of the intervention. For ethical reasons, inert placebos are often used, and this increases the chances that an active drug will be unmasked simply because it is active, but not because it is effective.

Worrall, Howick, and others are helpfully reminding us that the reliability of evidence-based medicine does not and should not depend upon a rigid and unreflective adherence to double-masked randomized controlled trial methodology. There is not an algorithm of steps that must be followed in order to do evidence-based medicine. Instead, there are specific recommendations for the avoidance of bias in particular kinds of situations and a general recommendation to exercise good "judgment" overall (see especially Rawlins 2008). The standard hierarchy of evidence, which puts double-masked randomized controlled trials at the top, is most useful for assessing interventions with small, variable, and subjectively measured effects, but even in these cases the hierarchy should not be applied too rigidly. The strength of evidence for an intervention depends in part on the likelihood of bias, and the likelihood of bias is determined by both trial methodology and the particular study details.

In his Harveian Oration, Sir Michael Rawlins stated that:

The notion that evidence can be reliably placed in hierarchies is illusory . . . Decision makers need to assess and appraise all the available evidence irrespective of whether it has been derived from randomized controlled trials or observational studies; and the strengths and weaknesses of each need to be understood if reasonable and reliable conclusions are to be drawn. (Rawlins 2008: 2152)

Rawlins was the Chairman of NICE from 1999–2012 and a longtime advocate of evidence-based medicine. Worrall and Howick would probably agree with this assessment.[12] Many—perhaps most—evidence-based medicine practitioners are, in practice, more rigid about the evidence hierarchy than Rawlins seems to be. This chapter (and the work of Worrall, Howick, and others) is not attacking a straw-person position.

What happens to the process of systematic review if we allow the evidence hierarchy to be more nuanced? One consequence is that we may

---

[12] In fact, John Worrall said in conversation to me that if all evidence-based medicine practitioners were like Rawlins, he would not need to do his philosophical work on evidence-based medicine.

not be able to do a systematic review at all, because of the difficulty of devising a nuanced and also impartial hierarchy of evidence. There is a shaky line between contextual and biased determinations. The danger is that departure from a rigid standard "for good reasons" may open the door to any number of rationalized (rather than "rational") departures from that standard. We may lose the actual and/or the perceived objectivity of systematic review, both of which are necessary. We may also lose the perceived analytic rigor of systematic review, which contributes to its perceived objectivity.

There is a tension between the need for standards to be contextual (adapted to context rather than general) and the need for standards to be impartial (which is achieved with generality and abstraction from concrete cases). Contextuality and partiality (meaning bias) are also easily confused by both participants and observers. Those who advocate a more contextual approach need to provide some reassurance that their standards are sufficiently impartial. Objectivity (at least partly shown by inter-rater reliability) is also important. Standardization is a combination of impartiality and objectivity. The CONSORT statement is an attempt to standardize the evaluation of randomized controlled trials, although it does not help us with ranking decisions when we have a variety of kinds of clinical trials. For the latter, the GRADE Working Group is attempting to produce a rating system with a more nuanced evidence hierarchy, acknowledging that the result does not have as much inter-rater reliability as that produced by earlier more rigid hierarchies.[13]

## 6.3   External Validity: Dependence on Background Knowledge and Judgment

Doing evidence-based medicine well requires background knowledge and judgment. This was not adequately appreciated in the early days of evidence-based medicine. In Chapter 5, I showed the importance of background scientific knowledge for inventing interventions

---

[13] Advocates of the GRADE system suggest that the details of each judgment of trial quality, and not the overall "grade" received, matter most for its goal of standardization of trial assessment. That is, they acknowledge a subjectivity in overall assessments of trial quality. See the presentation by Dr. Holger Schunemann to the CDC, which is on the GRADE website at <http://www.gradeworkinggroup.org/toolbox/index.htm>, accessed October 20, 2014.

and designing clinical trials. In a series of articles (2007a, 2007b, 2009, 2012, Cartwright & Munro 2010), Nancy Cartwright has argued that the proper application of trial results to clinical decision making also requires a good deal of background knowledge and judgment. She bemoans the "vanity of rigor" (2007a) which afflicts some advocates of evidence-based research, leading them to overlook the presuppositions in their applications of formal methods. Many other philosophers, epidemiologists, and clinicians share Cartwright's concerns about external validity; I will discuss them and their additional concerns in this section and in Section 6.4.

Cartwright's focus is not on the internal validity of randomized controlled trials (this was Worrall and Howick's focus) but on their external validity.[14] She points out that external validity is dependent on the similarity of the population and context of the trial participants to the population and context of those targeted for intervention. She uses primarily non-medical examples. For example, she cites the failure of the California program to reduce class sizes in schools,[15] which was based on the success of a randomized controlled trial in Tennessee, as due to differences in the implementation of the program and in background conditions (Cartwright 2009). She does not give a medical example of a failure of external validity, although she gives one of possible failure: prophylactic antibiotic treatment of children with HIV in developing countries. UNAIDS and UNICEF 2005 treatment recommendations were based on the results of a 2004 randomized controlled trial in Zaire. Cartwright is concerned that the Zaire results will not generalize to resource-poor settings across other countries in Sub-Saharan Africa (Cartwright 2007b).[16]

---

[14] Cartwright does not use the term "external validity" because (as I understand her views) she thinks that so much work needs to go into figuring out whether and how a study result might hold elsewhere that she prefers to summarize that work by using the language of contexts and capacities. I agree that external validity is not a simple or general property of a trial, but I still think that the term is useful.

[15] The claim that the program reducing class size in California schools was a failure is controversial. I am not agreeing with Cartwright, but giving an example of a case in which she thinks that the results of randomized controlled trials do not simply generalize.

[16] I am following this example and so far there is no evidence for lack of external validity. Other countries in which prophylactic antibiotics for children born to HIV-infected mothers are being used successfully are Zambia and South Africa. See Chintu et al. (2004), Coutsoudis et al. (2005), Prendergast et al. (2011).

Concern about external validity is reasonable and there are some classic medical examples of lack of external validity. For example, some recommendations for the treatment of heart disease, developed after trials of men only, do not apply to women (Spencer & Wingate 2005). Many of the examples come from a history of challenges to randomized controlled trials on the grounds that they have excluded certain groups from participation (e.g. women, pregnant women, the elderly, children) yet are used for general health recommendations (see, for example, Rogers 2004). Until 1993, the FDA excluded women of childbearing age from participating in early (Phase I and II) drug treatment trials, and this reduced their enrollment in Phase III trials (usually randomized controlled trials). These days, women are less likely to be excluded because the NIH and other granting organizations require their participation in almost all clinical trials, but other exclusions, such as those based on age and comorbidities, remain. Cartwright expresses the concern about external validity in its most general form: how do we know that an intervention will have the same effects in a community that is different (in known and unknown respects) from the trial community?

In her 2010 Philosophy of Science Association Presidential Address (2012) Cartwright describes four conditions that need to be met for us to have confidence that the results of a trial will project successfully to a new setting. These are: (1) "Roman Laws," or laws that are general enough to cover both the trial situation and the new setting; (2) "the right support team," or presence of all necessary background conditions from the trial situation in the new setting; (3) "straight sturdy ladders," for climbing up and down levels of abstraction (what this means need not be discussed here); and (4) "unbroken bridges" or no interfering conditions. These four conditions demand considerable domain knowledge, i.e. knowledge of the particular causal interactions that the intervention relies upon. The result is, at best, confidence (not certainty) that an intervention will work in a specific new setting.

Cartwright is skeptical that interventions will work in all contexts and thereby have general external validity. She uses examples from education, economics, and international development—not medicine—to show lack of external validity. In general, her examples show failures of interventions to generalize, often because of cultural differences between populations.

In fact we already know a good deal about reasons for lack of external validity in medical trials. Rawlins (2008) summarizes the known issues in terms of patient variability (age, sex, severity of disease, risk factors, comorbidities, ethnicity, socioeconomic status),[17] treatment variability (dose, timing of administration, duration of therapy, other medications taken), and setting (quality of care). In order to address these issues, knowledge of their importance in particular cases is needed. Domain expertise is essential for projection, as Nelson Goodman argued long ago (1955). Cartwright's four conditions, described above, have a role here as criteria for assessing external validity.

It should also be noted that this discussion of external validity applies broadly to experimental and evidential reasoning and not specifically to trial methodology, even randomized controlled trial methodology. The discussion is especially pertinent to randomized controlled trials, however, because randomized controlled trials are generally held in such high epistemic regard. It is also especially pertinent to randomized controlled trials because they are sometimes thought to be wholly governed by their own methodology (such as CONSORT standards). As argued in Chapter 5, evidence-based medicine methodology (which includes the methodology of randomized controlled trials) is not a complete method of inquiry, and its formulation and application require causal reasoning of the kind that Cartwright's four conditions cover.

Another approach to addressing questions about external validity, this time from the methodologists of social science Ray Pawson and Nick Tilley (1997) rather than philosophers, is what is termed "realist evidence synthesis" (or "realist review," or "realist evaluation"). Instead of Cartwright's four conditions for projectibility, realist synthesis uses a context-sensitive mechanistic account of the intervention to supplement traditional evidence-based reviews such as Cochrane reports (Greenhalgh et al. 2011). The objectivity of realist review lies not in following explicit procedures (which, on traditional interpretations, grounds evidence-based medicine) but in expert consensus—yet another example of the irony that evidence-based practice ends up appealing, at crucial junctures, to expert judgment. In looking at the details, Cartwright's

---

[17] I have left out "comparative effectiveness," which Rawlins includes, because I don't think it affects external validity (although it may affect ultimate treatment recommendations).

account and realist evidence review say much that is similar about the conditions needed for confidence that a successful intervention will be successful in a new setting.

## 6.4  External Validity Continued: On Biological Variability

One of the most common criticisms of evidence-based medicine, and especially of evidence-based clinical guidelines, is that generalities and rules don't always apply to individuals, who may have special features or circumstances. A common way of putting this is to say that "statistics don't apply to individuals." This is a form of failed external validity (as discussed in Section 6.3) if what was discovered for a test population does not project to a particular new circumstance with a new patient. Cartwright's four conditions (and the techniques of realist review) are intended to help with projection to particular individuals. They require domain expertise, not only familiarity with the evidence base. This domain expertise can be the contribution of the clinician. Indeed, in the classic restatement of the definition of evidence-based medicine David Sackett explicitly includes such "clinical expertise":

Evidence-based medicine is the conscientious, explicit, and judicious use of current best evidence in making decisions about the care of individual patients. The practice of evidence-based medicine means integrating individual clinical expertise with the best available external clinical evidence from systematic research. (Sackett et al. 1996: 71)

This statement is a carefully crafted backtracking from the 1992 original statement of evidence-based medicine, which explicitly advocated "de-emphasizing" what it called "unsystematic clinical experience" (Evidence-Based Medicine Working Group 1992). Now the individual clinical expertise of the physician is recognized, indeed respected, as an important part of the process.

According to traditional logical empiricist views of the nature of science, scientific generalizations or laws apply straightforwardly to their instances, so the clinician should need nothing more than deductive logic and a few simple observational facts (the "antecedent conditions") to figure out what to do in a particular clinical situation. Probably this is the way that champions of evidence-based medicine thought about the situation at first or when they were thinking abstractly. In practice, we discover that medical knowledge is less general than are the idealized laws of nature[18] as soon as the clinician asks the question "Is my patient sufficiently similar to the trial subjects for the results of the trial to apply to her?" The patient might be frailer than the trial patients, and/or much older, or be of a different gender, or have significant comorbidities, or lack the requisite social support. Part of clinical expertise is having the ability to answer this question, perhaps drawing on general pathophysiological and psychosocial knowledge as well as knowledge of the particular patient. We lack a more ideal generality because of trial design as well as because of biological and cultural variability and complexity. Trial design often has goals, such as the desire to show therapeutic effects of a new intervention, that conflict with a more general study of the intervention's effectiveness. Hence elderly patients or patients with comorbidities are often excluded from clinical trials when they are expected to do less well with the new therapy. Even when such goals are absent, trial populations can differ importantly from patient populations due to geography, access to health care, culture, and so forth.

In a paper "In defense of expert opinion," which is an early criticism of evidence-based medicine, Mark Tonelli writes:

> The randomized controlled trial must obscure the effect of individual variability, resulting in knowledge that is referable only to an "average" patient. The "average" patient, of course, does not exist. (1999: 1190)

Tonelli describes the situation as one of an "epistemic gap . . . between the results of clinical research and the care of individual patients" (1999: 1188) and argues that expert clinician opinion is needed to bridge that gap. In a contemporaneous paper (Tonelli 1998), he points out that the gap is also an ethical gap, since action is being contemplated in the context of the patient values and preferences and the physician's ethical standards. Tonelli also claims that clinical judgment contains a "tacit element" that cannot be explicitly stated or captured by a decision model. For Tonelli, such partly tacit "clinical expertise" is necessary to close the epistemic and ethical gaps between the results of medical research and the care of the individual patient. I'm going to try to proceed without

---

[18] Cartwright (1983) has argued for some time that even in physics there are no general laws.

appeal to indescribable elements; even tacit knowledge can be examined, observed, described, and discussed.

Others have made similar or related remarks about the relevance of research conclusions to individual patients. Stephen Jay Gould's essay, "The median isn't the message" (1985), has been widely cited and used to argue that "statistics do not apply to individuals." Gould was diagnosed with abdominal mesothelioma in 1982, a disease which then had a median life expectancy of eight months. Gould argues that it does not follow that he should expect to live "about eight months." He argues for a different interpretation of the statistics. Gould reflects that the life expectancy curve for abdominal mesothelioma is right skewed, with a long tail representing patients who live much longer than eight months. And he provides reasons to think that his case will be in the long tail, because he has characteristics (youth, optimism, access to new treatments) that are associated with a longer lifespan. Gould does not go so far as to say that "statistics do not apply to individuals"; rather, he says that we need to be careful about how we interpret the statistics, and remember that they are mathematical constructs describing a highly variable reality. For Gould, our biological essence is found in our variability, as he says, "Variation is the hard reality . . . means and medians are the abstractions" (1985: 41).

The term "cookbook medicine," which is invariably used pejoratively, describes the practice of unreflectively using general clinical guidelines to treat individual patients. "Cookbook medicine" is derided by those who use the term because they think that general rules are insufficiently nuanced for the treatment of individual patients. The same point is being made (this time about clinical guidelines rather than about the results of clinical trials): patients and their situations are biologically highly variable, and general knowledge and guidelines do not straightforwardly apply. It requires clinical expertise to apply these generalities to particular clinical cases.

There is a danger in departing from established guidelines: some studies have shown that physicians find "exceptions to the rule" more often than appropriate, and that their results can be worse than would be obtained by an unreflective obedience to the rule. Pathophysiological reasoning is particularly susceptible to cognitive biases such as availability bias, base rate fallacy, resemblance, and anchoring by past salient cases (see, for example, Elstein 1999, Eva & Norman 2005, Klein 2005, Groopman 2008).

Gould is correct in saying that it is biologically variable individuals, rather than statistical averages, that exist in the world. He is also correct to point out that it does not follow from the fact that the median life expectancy is eight months that Gould should expect to live about eight months. (Indeed, why expect to live the median rather than, for example, the mean or the mode?) He is, however, quite biased in his attempts to tailor the prognosis to his own circumstances. What actually follows from the median life expectancy is that Gould's chances of living longer than eight months are equal to his chances of living less than eight months: not a comforting thought! Even if youth, optimism, and access to new treatments predicted greater longevity (and neither we nor Gould have good supportive data for this), they would probably not predict the length of life (20 years) that Gould ended up having. Gould's article is popular because it provides hope, not because it is a well-reasoned prediction of his expected longevity. In more than one way, Gould got lucky. Had he not lived so long, his reasoning about probabilities would not have been memorable.

Clinical skills include applying abstract or general results (such as statistical results and guidelines) to individuals whose characteristics may or may not have been represented in trial populations. This is an applied scientific skill, familiar from other domains such as weather forecasting, in which we often take results from one context (e.g. hurricanes in the southern US) and apply them in another context (e.g. hurricanes in the northern US). Pathophysiological, or causal, reasoning is the method used, while remembering its vulnerability to error because we typically have incomplete knowledge of causal factors in both contexts.

Some (perhaps Tonelli 1999 and Braude 2012, and certainly Montgomery 2006) would also argue that "statistics do not apply to individuals" for another reason: because we are dealing with *human beings* (rather than with animals, plants, or inanimate matter). They argue that evidence-based medicine and pathophysiological reasoning, together, do not fully capture what is needed in what we now call "personalized medicine."[19] They think that there is something more to clinical skill than the ability

---

[19] "Personalized medicine" is a term first coined in relation to the hope that new genetic therapies would be tailored to individual mutations. The term has expanded to include all ways of tailoring medical knowledge in application to specific individuals and to signify that "cookbook medicine" is not being used. I think that the ambiguous usage confuses more than it enlightens.

to go from knowledge generalities to specific applications (which is an applied scientific skill). While Tonelli appeals to "tacit" knowledge and Braude to "intuition," Montgomery describes the "je ne sais quoi" of the good clinician. My approach is to try to understand this further: is the "tacit" knowledge of the clinician the same as the "tacit" knowledge of the scientist in other applied contexts? In other words, are "intuition" and "je ne sais quoi" skills that depend on the clinician's general scientific expertise—as Michael Polanyi (1958) would argue—and/or skills such as "people skills" (sometimes referred to in medical schools as "soft skills")? I think that the answer to this question is "probably both, but definitely the latter." Some of what is going on here is a commitment to the personhood, and not only the biological individuality of the patient. In addition, I think there is an effort to retain the professional autonomy of the physician. These issues will be discussed in Chapter 8.

## 6.5 Empirical Studies of the Reliability of Evidence-Based Medicine

How reliable is evidence-based medicine in practice? Randomized controlled trials and meta-analyses generate claims that are usually expressed as probabilities of getting specified results on the null hypothesis. The convention is that randomized controlled trials are expected to deliver significance levels of less than or equal to 0.05, and meta-analyses offer greater significance than that. Such significance levels do not, however, tell us about the probability that the null hypothesis (or any other hypothesis) is true. (To get these probabilities requires knowledge of the prior probabilities of the various hypotheses.) All that can be inferred from a trial with positive results and a p-value of less than or equal to 0.05 is that we have some evidence against the null hypothesis (Goodman 2008). And the more successful replications of the study, the more evidence against the null hypothesis we have.

Ioannidis (2005) did a study of attempted replications of 59 highly cited original research studies. Less than half (44 percent) were replicated with the same results; 16 percent of were contradicted by attempted replications and 16 percent of the replications found the effect to be smaller than in the original study; the rest (24 percent) were neither repeated nor challenged. And LeLorier et al. (1997) found that 35 percent of the time the outcomes of randomized controlled trials are not predicted accurately by previous meta-analysis (which combines the results of several randomized controlled trials).

This is a large and partly unexplained rate of failure to replicate. (It is especially surprising for those who misinterpret p-values, which is common.[20]) Some suggest that factors such as publication bias and time to publication bias are partly responsible for the worse-than-expected track record of randomized controlled trials, systematic reviews, and meta-analyses. Publication bias occurs when studies with null or negative results[21] are not written up or not accepted for publication because they are (incorrectly) thought to be of less scientific significance. A statistical technique called "funnel plots" can also be used to check for publication bias. Time to publication bias is a more recently discovered phenomenon: trials with null or negative results, even when they are published, take much longer than trials with positive results (six to eight years for null or negative results compared with four to five years for positive results) (Hopewell et al. 2007). Trial registries requiring results reporting may eventually help researchers correct for both publication bias and time to publication bias. Over the past ten years, trial registries have been created nationally and internationally that require registration of trials and their measured endpoints prior to beginning data collection as well as reporting of the final data. In the US, this is a joint project of the FDA and the NIH, called ClinicalTrials.gov. It is too early to know whether or not this will help correct for publication bias.[22]

Pharmaceutical funding biases are major contributors to failures of replication. An astonishing finding is that studies funded by pharmaceutical companies have three or four times the probability of studies not funded by pharmaceutical companies of showing effectiveness of an intervention (Als-Nielsen et al. 2003, Bekelman et al. 2003, Yank et al. 2007, Lexchin et al. 2003). The additional bias created by pharmaceutical

[20] I was one of those people (Solomon 2011b), until Ken Bond helped me clarify my confusions.

[21] In this context, a positive result is one in which the experimental arm of the trial is more effective, a null result is one in which both arms are equally effective, and a negative result is one in which the control arm is more effective.

[22] In my own inquiries into the fate of recent clinical trials for treatment of cystic fibrosis, I find that recently (within the past five years or so) failed clinical trials (e.g. of Denufosol) have not yet reported their data to ClinicalTrials.gov.

funding is not fully understood, especially since many of these trials are properly randomized and double masked and satisfy rigorous method-ological criteria. Some (for example Angell 2005, Sismondo 2008) sug-gest that pharmaceutical companies deliberately select a weak control arm, for example by selecting a low dose of the standard treatment, giv-ing the new drug a greater chance of relative success. There can also be biases that enter into the analysis of data, particularly when endpoints are not specified in advance. And pharmaceutical companies contribute to publication bias, since they are not interested in having their failed trials publicized. It is hoped that clinical trials registries will help cor-rect for both hindsight manipulation of endpoints and publication bias. However, at present we are a long way from correcting for all the biases created by pharmaceutical funding. Disclosure of funding source, which is increasingly required, is helpful for the evaluation of results, although there are no formal procedures for how to take such conflicts of inter-est into account.[23] Moreover, information about conflicts of interest is typically lost in systematic review and meta-analysis.

Since the results of randomized controlled trials are so fallible, it is worth asking the question whether other kinds of clinical trials, further down the evidence hierarchy, are even less reliable. One might think that since the further down the evidence hierarchy the more *possible* kinds of bias, then the further down the evidence hierarchy the more *actual* quantity of bias and unreliability. But this is an empirical question. Jan Vandenbroucke (2011: 7020) remarks wryly that "empirical proof that observational studies of treatment are wildly off the mark has been sur-prisingly elusive." He cites four meta-analyses (Benson & Hartz 2000, Concato et al. 2000, MacLehose et al. 2000, Ioannidis et al. 2001) con-trasting the results of randomized controlled trials and observational studies which find "no systematic differences." The matter is still con-troversial, but an article by Ian Shrier et al. (2007) strongly argues for the inclusion of observational studies in systematic reviews and meta-analyses, and a new Cochrane review finds that "there is little evi-dence for significant effect estimate differences between observational studies and RCTs [randomized controlled trials]" (Anglemyer et al. 2014: 3). Vandenbroucke (2009) also points out that observational stud-ies are superior to randomized controlled trials for the determination of

---

[23] I will suggest a formal procedure in Section 6.7.

adverse effects. The oft-cited example of the difference between observational studies and randomized controlled trial studies of hormone replacement therapy and cardiac health (in which randomized controlled trials look like stronger methodology) has been analyzed by Vandenbroucke (2009) and found to be due to trial differences other than randomization.

This result corroborates the position of early AIDS activists, who argued against the imposition of randomized controlled trials for AZT on both ethical and epistemic grounds (Epstein 1996). They argued that such trials are morally objectionable in that they deprive the individuals in the placebo arm of the only chance for a cure. And they argued that such trials are epistemically unnecessary because a randomized controlled trial is not the only way to discern the effectiveness of anti-retroviral drugs. This claim has, in hindsight, proved correct as a combination of historically controlled trials[24] and laboratory studies have provided the knowledge of dramatically effective antiretrovirals in clinical use today. These days of course, no one needs to get a placebo treatment for HIV-AIDS, and randomized controlled trials can continue to detect small improvements of protocol without such strenuous moral objections.

Finally, it is reasonable to ask evidence-based medicine to evaluate itself using its own standards of evaluation. This involves going further than showing that specific interventions improve outcomes. It needs to be shown empirically that the general use of evidence-based medicine in clinical decision making results in improved outcomes for patients. That is, it needs to be shown that using systematic evidence reviews, and the clinical guidelines based on them in patient care, produces better results. In theory, all other things being equal, we would expect improved outcomes. But we do not know whether or not all other things are equal, and it is likely that there are systematic differences (other than their use of evidence-based medicine) between physicians and institutions practicing evidence-based medicine and those not practicing it. What matters here is not theory but practice, and no one has yet designed or carried out a study to test this (Charlton & Miles 1998, Cohen et al. 2004, Straus & McAlister 2000).

---

[24] "Natural experiment" observational trials would have been less desirable here because of the likely differences between the treated and untreated groups.

## 6.6 Choices for Aggregating Complex Evidence

Systematic review involves a thorough search of the literature, an assessment and grading of the evidence, meta-analysis where possible, and finally judgments about the aggregate of evidence. Bias can enter in a multitude of ways, from literature search techniques that miss some relevant studies, to incorrect assessments of the evidence, to arbitrary assessments of the grade of evidence (especially when there is more than one grading system in use), to inappropriate decisions during meta-analysis.

Standards, techniques, and performance for systematic review and meta-analysis are improving all the time, in that strategies are invented to avoid or correct for known biases. Yet there are at least two reasons for thinking that we are not on the way to a general specification of an algorithmic method. First, domain expertise is needed to do systematic reviews and meta-analyses. The typical team producing an evidence report is composed of both methodologists (with statistical training) and domain experts (who know the area of medicine under review). Second, there are several possible classifications of strength of evidence, as discussed in Chapter 5. In 2002, the AHRQ reported 40 systems of rating in use, six of them within its own network of evidence-based practice centers (AHRQ 2002). Although these classifications have some overlap (e.g. they all rank randomized controlled trials high and anecdotal evidence low), there is some variation that can make a difference in cases where the evidence is "close," for example on the question of the benefits of routine-screening mammography for women in their forties (see Chapter 9 for a full discussion of this example). And, all too often these days, the evidence turns out to be "close" because new drugs and treatment protocols typically offer at best small improvements in efficacy.

The GRADE Working Group, established in 2000, is attempting to reach expert consensus on one system of rating the quality and strength of evidence (Brozek et al. 2009a, Brozek et al. 2009b, Guyatt et al. 2008, Kunz et al. 2008). As mentioned in Chapter 5, this is an ironic (although telling) development, since evidence-based medicine is generally intended to replace group judgment methods. Some group judgment is indispensable.

Evidence-based medicine is not a general, complete, all-purpose, algorithmic, and infallible methodology of science. This finding should not be too surprising to philosophers of science, since traditional assumptions about the generality of scientific method have been extensively challenged (Bechtel & Richardson 1993, Fine 1986, Keller 1985, Kuhn 1962, Laudan 1977). Evidence-based medicine enthusiasts are beginning to acknowledge this sensible moderation of their views, as the Harveian Oration by Sir Michael Rawlins shows. In this paper, Rawlins argues that "the notion that evidence can be placed reliably in hierarchies is illusory," (2008: 2152) the findings of randomized controlled trials should be "extrapolated with caution," (2008: 2157) and that "striking effects can be discerned without the need for randomized controlled trials" (2008: 2155).

Accepting these qualifications puts evidence-based medicine in the same normative category as other successful scientific methods (and methodologies). Successful methods are useful tools in the domains in which they work, but they do not work everywhere or always. Knowledge of the qualifications is important because it can help investigators use the methods more judiciously. Some evidence-based medicine advocates have made or still make inflated claims about evidence-based medicine, thinking of it as a universal and algorithmically applicable set of standards. This section and this chapter are a corrective to that kind of thinking without swinging to an opposite conclusion that is generally skeptical of evidence-based medicine.

## 6.7  Some Simple Recommendations

Trial registries may help produce more reliable research by helping to prevent manipulation of endpoints, publication bias, and time to publication bias. Once the trial registries are established, we can detect any improvement in reliability by repeating earlier studies (such as Als-Nielsen et al. 2003, Bekelman et al. 2003, Yank et al. 2007, Lexchin et al. 2003, Ioannidis 2005).

The most common suggestion for correcting for pharmaceutical company bias (a form of conflict of interest) is to recommend that pharmaceutical companies no longer be involved in research and that instead we have national or international testing funded by government research budgets. Such suggestions have been made by Sheldon Krimsky (2003), Marcia Angell (2005), and Sergio Sismondo (2008). The difficulty with these suggestions is that they show little chance of being implemented soon, especially in the US.

Here is an interim (and tentative) suggestion. Trials with conflicts of interest that result from ties to pharmaceutical companies should occupy a place in trial hierarchies that is *below* that of observational trials. This is because observational trials have turned out to be *more* reliable than expected. Since it looks like randomized controlled trials and observational trials have comparable validity when they are of high quality, it makes sense to rank pharmaceutically funded studies (which are three or more times likely to show benefit of the drug) at a level below. This means that they cannot be included in a meta-analysis, unless (1) observational trials are also included in meta-analyses and (2) the mathematics of meta-analyses is expanded to include a lesser weighting for data from less than highest-quality trials.

The bias in pharmaceutical company funded studies can be tracked over time by repeating the kinds of evaluation done by Als-Nielsen et al. (2003), Bekelman et al. (2003), Lexchin et al. (2003), and Yank et al. (2007) and seeing whether their reliability is improving. If they do improve, their place in the evidence hierarchy might be reconsidered.

## 6.8  Conclusions

Evidence-based medicine is a set of formal techniques for evaluating the effectiveness of clinical interventions. The techniques are powerful, especially when evaluating interventions that offer incremental advantages to current standards of care, and especially when the determination of success has subjective elements. For reasons that are not yet fully understood, evidence-based medicine techniques do not deliver the reliability that is expected from them. Results are compromised by significant differences between trials and clinical populations, other differences between trials and clinical practice, publication bias, time to publication bias, interests of funding organizations, and other unknown factors. Maintaining a strict evidence hierarchy makes little sense when the actual reliability of "gold standard" evidence is so much less than expected. I recommend a more instrumental or pragmatic approach to evidence-based medicine, in which any ranking of evidence is done by reference to the actual, rather than the theoretically expected, reliability of results.

# 7

# What Is Translational Medicine?

Ask ten people what translational research means and you're likely to get ten different answers. (Butler 2008: 841)

The creation of a redefined discipline of translational medicine will require the emergence of a new and vibrant community of dedicated scientists, collaborating to fill knowledge gaps and dissolve or circumvent barriers to improved clinical medicine. . . . A profound transition is required for the science of translational medicine. (Mission Statement of *Science Translational Medicine*)[1]

Translational efforts are as old as medicine. (Wehling 2008)

## 7.1 Introduction

Since around 2002, translational medicine has become a prominent new focus of granting organizations, journals, research institutions, conferences, and professional associations. For example, it features centrally in the 2004 NIH Roadmap for Medical Research and the subsequent funding of 60 Institutes for Translational Research and a National Center for Advancing Translational Sciences in the US. The UK National Institute for Health Research funds 11 centers for translational research. The European Advanced Translational Research InfraStructure in Medicine supports centers for translational medicine in eight countries. At the

---

[1] From the "Mission Statement" of *Science Translational Medicine*, and in part a quote from Elias Zerhouni. Zerhouni was director of the NIH from 2002 to 2008, which includes the period during which the NIH Roadmap for Medical Research, advocating translational research, was released and implemented: <http://stm.sciencemag.org/site/about/mission.xhtml>, accessed July 24, 2014.

time of writing, seven general journals devoted to translational medicine and many more specialized journals have been established.[2] The
US-based Association for Clinical and Translational Science (ACTS)
was founded in 2009 with a mission "to advance research and education
in clinical and translational science so as to improve human health"[3]
and the European Society for Translational Medicine was founded in
2013. Masters- and doctoral-level programs in translational medicine
are widely offered. A well-known course on translational medicine,
Demystifying Medicine, created and directed by Irwin Arias ("the
father of translational medicine") in 2002, is offered at NIH to PhD scientists and currently enrolls about 800 participants annually.[4]

Translational medicine, like other new medical methods such as consensus conferences, evidence-based medicine, and narrative medicine,
promises to be *transformative* for medical knowledge and practice. Yet
unlike these other new methods it has particularly sketchy and varied definitions. The quotations used as epigraphs for this chapter, taken together,
suggest that translational medicine may not even be a coherent concept.
The most common response I have received from those who hear that I am
writing about translation medicine is: "What *is* translational medicine?"
(Hence the title of this chapter.) The purpose of this chapter is to develop
some clarity about the meaning(s) of translational medicine and then identify its role, among other methods, in biomedical research and practice.

I will begin with a careful look at recent and current usages. I will
argue that translational medicine needs to be understood historically,
philosophically, and sociologically, as a response to the combined shortfalls of consensus conferences, evidence-based medicine, and basic science research in the 1990s.[5] That is, it is a response to recent shortfalls in

---

[2] New general journals are the *Journal of Translational Medicine* (2003–), *Translational Research* (2006–), *Clinical and Translational Science* (2008–), *Science Translational Medicine* (2009–), *American Journal of Translational Research* (2009–), *Drug Delivery and Translational Research* (2011–), and *Clinical and Translational Medicine* (2012–). Many more specialized journals, such as the *Journal of Cardiovascular Translational Research* (2008–), *Experimental and Translational Stroke Medicine* (2009–), *Translational Stroke Research* (2010–), and *Translational Behavioral Medicine* (2011–) have also been established. Most of these journals are new; in a few cases they involved renaming established journals.

[3] See <http://www.ctssociety.org/?page=Background>, accessed July 24, 2014.

[4] The most recent version of the course is at <http://demystifyingmedicine.od.nih.gov/>, accessed July 31, 2014.

[5] Translational medicine also emerged in a particular political context—the NIH under Elias Zerhouni in the early 2000s—although it is now an international movement. I'm sure that there is more to say about the emergence of translational medicine than I have been able to uncover.

both methods and results, and a phenomenon of the first decades of the twenty-first century.

## 7.2 Definitions of Translational Medicine

An early use of the term "translational research"[6] defines it as "walking the bridge between idea and cure" with the expectation that the bridge between these two worlds is difficult but necessary to build (Chabner et al. 1998: 4211).[7] A similar definition is given a few years later for the term "translational medicine" in the phrase "from bench to bedside," and is explicated as follows:

A basic laboratory discovery becomes applicable to the diagnosis, treatment or prevention of a specific disease and is brought forth by either a physician-scientist who works at the interface between the research laboratory and patient care or by a team of basic and clinical science investigators. (Pizzo 2002)

In other words, translational medicine "translates" pure science knowledge into effective health applications. (What is meant by "translation" will be discussed in due course.) An equivalent—and similarly alliterative—common explanatory phrase is the "Petri dish to people" definition of translational medicine (Pardridge 2003).

This "bench to bedside" phrase is the most common explication of "translational medicine," and it is often elaborated by making it reflexive. Early on, Francesco Marincola (editor of the *Journal of Translational Medicine*) insisted that translational medicine should be thought of as a "two way road: Bench to Bedside and Bedside to Bench." In particular, Marincola urged that the results of Phase I clinical trials be used as clinical feedback for the design of further Phase I or Phase II trials (Marincola 2003). Marincola's suggestion has been widely reiterated in the definitions of others (e.g. by Horig & Pullman 2004, Mankoff et al. 2004, Woolf 2008). In Section 7.4 I will explore what is at stake in the insistence that translational medicine is a two-way, rather than a one-way, road.

The term "translational medicine" is *also* widely used to describe the work of moving new therapies from the research context to everyday

---

[6] The term "translational research" emerged in the late 1990s; the term "translational medicine" in the early 2000s.

[7] The mixed metaphors here—crossing the bridge versus building it—are typical of the discourse of translational medicine.

clinical usage (Woolf 2008). For example, several of the presentations at the meetings Translational Science 2012 and Translational Science 2014 were devoted to the move from research to everyday practice.[8] The journal *Translational Behavioral Medicine* is an example of an entire journal focused on this meaning of "translational medicine." Woolf adopts the convention of calling the move from clinical research to regular clinical practice "T2," thus distinguishing it from the "bench to bedside (and back)" usage, which is called T1.[9] I will stay with this usage of T1 and T2, even though there are proposals for more fine-grained distinctions (Khoury et al. 2007, Schully et al. 2011). Most of the funding initiatives, both nationally and internationally, are for T1 projects. Schully (2011) estimates that less than 2 percent of the research funds and 0.6 percent of the literature is T2 research. Notably, however, the AHRQ Translating Research Into Practice (TRIP) initiative focuses on T2 projects.

In addition, the term "translational medicine" has been used to describe pretty much everything on the medical research horizon, as this quotation from Pizzo (2002) illustrates:

Translational medicine may also refer to the wider spectrum of patient-oriented research that embraces innovations in technology and biomedical devices as well as the study of new therapies in clinical trials. It also includes epidemiological and health-outcomes research and behavioral studies that can be brought to the bedside or ambulatory setting.

Similarly, the ACTS says that its "unifying theme is a commitment to apply scientific methodologies to address a health need,"[10] which does not distinguish translational medicine from the past 20 years of evidence-based medicine initiatives, or even from the over 150-year-old tradition of scientific medicine.

Another definition is that of Matthew Liang, who takes translational medicine to include "translating the disease described in textbooks to illness." By "illness," Liang means the experience of patients

[8] See <http://www.translationalsciencemeeting.org/sites/default/files/Final%20 Program-FINAL/pdf>, accessed September 1, 2012, and <http://c.ymcdn.com/sites/www. actscience.org/resource/resmgr/Translational_Science_2014_Meeting/ACTS_2014_ Final_Program.pdf>, accessed July 24, 2014.

[9] The terms "T1" and "T2" originated at NIH around 2005 (see Kerner 2006). I am grateful to an anonymous referee for this reference.

[10] <http://www.actscience.org/?page=Background> accessed October 30, 2014.

(Liang 2003: 721).[11] Even bioethicists have jumped on the "translational medicine" bandwagon; a 2009 conference was titled Books to Bedside: Translational Work in the Medical Humanities[12] and featured the usual range of work in the medical humanities. (There was no special focus on the "translational" work of interpreting books for use in clinical situations.)

It is not helpful to call everything that might interest us "translational medicine"; concepts have purchase only when we are willing to exclude some things from their scope. I have a sense that there is some jumping on the translational medicine bandwagon, to include any projects where the word "translation" can be made to fit, however awkwardly!

I am not looking for necessary and sufficient conditions for the use of the term "translational medicine." This kind of precision is inappropriate for most human practices (despite the wishes of some analytic philosophers). I am looking for meaningful content. Thus I look at both the stated definitions and the usage of the term.

This chapter will focus on the two core usages of "translational medicine," namely T1, i.e. applied research from bench to bedside (and back), and T2, i.e. moving successful new therapies from research to clinical contexts. Some of the wider usages of the term are, in my view, more opportunistic than substantive and I see no reason to develop them. Sections 7.3 to 7.8 explore T1, which is much more prominent than T2. T2 will be discussed in Section 7.9.

## 7.3   Example of a Translational Medicine Success

Here is an example of a recent translational medicine success.[13] Emma Whitehead, at age 7, had treatment resistant acute lymphocytic leukemia and was near death in October 2011. She was the first child treated with

---

[11] Descriptions of disease in textbooks are not really "translated." Rather, the patient's experience is gathered (in interviews and/or memoirs). So this work actually is a poor fit for the metaphor of translation.

[12] See <http://bioethics.northwestern.edu/events/b2b_schedule.html>, accessed May 3, 2009. (Unfortunately the program is no longer posted online; I would be happy to send a copy to whomever is interested in seeing it.)

[13] See <http://www.chop.edu/service/oncology/patient-stories-cancer/leukemia-story-emily.html>, accessed October 20, 2014. Children's Hospital of Philadelphia (CHOP) does not explicitly advertise this case as an example of "translational medicine";

the CTL019 Protocol devised by University of Pennsylvania research-ers, who developed an in vitro process for genetically modifying T cells with material inserted by a modified HIV virus. The idea was to get the T cells to recognize and then destroy cancerous B cells. Emma Whitehead went into remission quickly and has been well ever since. Indeed, 18 out of 27 patients in early trials are in long-term remissions of their dis-ease. What makes this trial "translational" is that it was developed in the lab by reasoning about pathophysiological process. Theoretically, it should have worked (and it did work), but many therapies that theoreti-cally should work have not worked (recall some of the drugs developed for cystic fibrosis and discussed in Chapter 5). The therapy worked for Emma Whitehead and for 17 others, but it did not work for all those with treatment resistant acute lymphocytic leukemia who received the proto-col treatment.

Important follow-up questions to ask are: what are the characteristics of the people that the CTL019 therapy worked for versus the people who had a temporary response or no response at all? Are their cancers differ-ent, or is some other aspect of their physiology different? Would their response have been better if the protocol for preparing the engineered T cells was adjusted in some way? Asking these questions is the "bed-side to bench" step, also part of translational medicine. The idea is to do more such Phase I and II research before the major investment of going to advanced (i.e. Phase III) clinical trials.[14]

Note that this example of translational medicine uses genetically engineered cells as therapy. Translational medicine includes any kind of diagnostic or therapeutic intervention, and so incorporates pharmaceut-icals, stem cell therapeutics, medical devices, vaccines, nanotechnology, genetic diagnosis, etc.

---

I selected it because it is well known regionally, in the public record, and fulfills T1 criteria. Interestingly, it is also an example of "personalized medicine" since the therapy—modified T cells—is produced for each patient from their own cells. CHOP does advertise this case as an example of personalized medicine. This is not surprising, since the website is aimed at prospective patients and their families, who are likelier to be attracted by the idea of personalized medicine than the idea of translational medicine.

[14]  As of the time of writing (October 2014), Phase III trials have not yet begun. One of the challenges to overcome before beginning these trials is to streamline the production of the modified T-cells. The drug company Novartis, as well as cancer funding agencies, are supporting this research.

## 7.4 The "Valley of Death"

I am not the first person to notice that the intellectual challenges of translational medicine (in T1) do not sound particularly novel. With the exception of some fortuitous discoveries in empiric medicine (such as the findings that willow bark relieves pain and that lithium is effective for treating bipolar disorder), we have always attempted to go from knowledge of basic science to attempts to intervene in human health (and vice versa). In a recent paper by Jane Maienschein et al. (2008: 44) there is good summary of this point, and some further reflections:

The idea of translating research into clinical applications is not new; indeed, it may be coextensive with the history of biomedical research. But what is supposedly different here is an explicit recognition that translation is not easy, not inevitable, not unidirectional, and, indeed, not happening. This recognition resulted in the attempt to re-engineer health research institutions and practices so as to facilitate the bench-to-bedside translation.

So Maienschein et al. claim that it is the recognition of the *difficulty* and *bidirectionality* of translational research that is new.

Ismael Kola and John Landis (2004), who are researchers employed in industry, find that only one in nine pharmaceutical compounds makes it through drug development to clinical approval and that "the vast majority of attrition" occurs late in development, during Phase IIb and III trials. Their assessment is based on the ten-year period 1991–2000, and covers drugs from the ten biggest drug companies. The 1990s were the years of the Human Genome Project, which were years of high optimism that the Human Genome Project was "a road map to health" (Wexler 1992: 240). Kola and Landis attribute drug failure to non-generalizable features of human trial subjects and to use of animal models that do not project to humans. They conclude that the rate of attrition of drugs in the clinical pipeline is "simply too high." Another typical reaction is that of NIH scientist Alan Schechter, who stated that "We are not seeing the breakthrough therapies that people can rightly expect" (quoted in Butler 2008: 840).

Researchers often refer to the situation as "the valley of death"[15] between basic research and clinical therapies. The valley of death is blamed on the "gap" between basic and applied research. Pharmaceutical

[15] I have not been able to find the origin of this phrase. Many sources say that the term comes from the pharmaceutical industry.

companies are sometimes blamed for avoiding the riskiness of drug development that might close the gap by, instead, developing "me too" drugs: drugs which are minor variants of already proven formulations. The idea, then, is that government funding of translational research will take up the slack and (to repeat these overworked metaphors), close the gap and build the bridge that will save us from falling into the valley of death.

In an important study, Despina Contopoulos-Ioannidis et al. (2003) have shown that only about a quarter of the findings of basic research get as far as a randomized trial (typically a Phase III clinical trial), and that those that get this far have taken a long time, up to 20 years. They also found that only a tenth of the findings of basic research entered routine clinical use in this 20-year period—a result that roughly agrees with the 1 in 9 figure of Kola & Landis (2004). Their conclusion is that "The gap between basic and clinical research needs to be narrowed" (Contopoulos-Ioannidis et al. 2003: 483).[16] This study is particularly interesting because it looks at basic research published between 1979 and 1983, and its fate 20 years later. Basic research published during 1979 and 1983 was of course neither part of the Human Genome Project nor stem cell research. As I said earlier in this section, going from basic research to clinical applications is not a new kind of challenge, nor is it unique to contemporary enterprises such as the Human Genome Project and stem cell research. It looks like research in the era of the Human Genome Project was neither faster nor slower at translating basic research into clinical applications than earlier research.

Most medical innovation proceeds in somewhat clumsy and unpredictable steps, with a combination of frustrating failure and serendipitous success, often with interdisciplinary teams, and with shifts from laboratory to patients and back again. Sometimes the research proceeds fairly quickly, and sometimes quite slowly. Here are two examples: the discovery of insulin and the development of penicillin.

During the discovery of insulin for the treatment of diabetes, Frederick Banting and Charles Best took approximately two years, with the help of Bertram Collip and J.J.R. Macleod, to overcome both technical and

---

[16] It is their conclusion, not the logical conclusion. It's entirely possible that the process of moving from basic to clinical research cannot be made much more efficient or successful.

institutional difficulties and produce a technology that worked and was commercially viable. Banting had the original idea of duct ligation (which played no role in the final technology), Best was his student assistant, and they began experiments on dogs in the summer of 1921 and then human trials in January 1922. Collip joined them to help isolate insulin, and Eli Lilly stepped in at a late stage to scale-up production. There were many setbacks and failures along the way. The eventual success was dramatic (so dramatic that it did not need formal clinical trials). Insulin became commercially available in September 1923. Macleod was Banting's supervisor, and advised him (both well and badly) throughout. Banting and Macleod won the Nobel Prize in 1923 and shared their awards with Best and Collip.[17]

The development of penicillin for medical use took more than a decade after its basic rediscovery by Alexander Fleming in 1928, requiring years of clinical, laboratory, and industry work from Howard Florey, Ernst Chain, Norman Heatley, and others. Fleming concluded early on, incorrectly, that penicillin would not survive long enough in the body to have a therapeutic effect. Florey, Chain, and others took over the purification of the drug and successfully used penicillin in mice in 1939, but at first were unsuccessful in human trials; they did not realize that the dosage they were using was too low. It took another four years to get the dosage right and the production stepped up by Merck and Company. Fleming, Florey, and Chain received the Nobel Prize for this work in 1945.

Going back and forth between laboratory work, animal models, and what is observed in clinical settings was typical in both examples; hence the reflexive "bench to bedside and back" is a good descriptor of the work done in the discovery of insulin and the development of penicillin. In view of these examples, perhaps Martin Wehling is right to say "Translational efforts are as old as medicine" (2008).

The details of these examples may, however, be better known to historians of medicine than to medical researchers. Medical researchers trained at the time of the Human Genome Project and its promise of therapies following quickly from basic research expected faster and easier results, in part because they believed their own hype.[18] In the

---

[17] My information about this case comes from Bliss (1982). I am grateful to John R. Clarke for giving me a copy of this book and discussing it with me.

[18] The effort to get funding for the Human Genome Project contributed to inflated claims about expected results. See, for example, Kevles & Hood (1992) and Dreger (2000).

1990s, genetic therapy was expected to deliver its promise and justify the large investments in the basic research of the Human Genome Project. However, early results were disappointing (see, for example, Lindee and Mueller 2011). Similar successes were expected and have not yet been delivered by stem cell research.

If going from basic research to clinical applications is not a new problem, and the rate of success has not significantly changed recently, why the new terminology and the new research initiatives? A cynical answer is that the new terminology is suggestive of new methodology and/or new knowledge and can help justify arguments for attention and resources, even though there is in fact no new methodology. I have actually been surprised by how *little* cynicism there is about translational medicine. With few exceptions,[19] translational medicine is taken with seriousness and sincerity by the scientific and medical research communities. I concur with this, and will argue that the new terminology of "translational medicine" is *more than* (although it may include) an advertising slogan for business as usual. There is substance to the translational initiative.

## 7.5  Metaphor in Translational Medicine

Although, as I will argue, there is substance to the translational initiative, that substance is somewhat shrouded in metaphor. I have already mentioned the "bridge" that "closes the gap" and leads us over the "valley of death." The word "translation" in "translational medicine" is obviously used metaphorically. There is no actual translation going on; translational medicine is not literally about meaning in symbolic language.[20] Yet the metaphor of translation is taken a long way, as far as talking about the importance of fluency in two "languages," namely basic research and clinical research. Really, translational medicine is not much like literary

---

[19] A notable exception is Gerald Weissman (2005), who argues that individual scientific curiosity and creativity are what is needed, rather than efforts to build buildings, facilitate interdisciplinarity, flatten the hierarchy, and publish preliminary findings. The fact that this argument is so rare leads me to conclude that most scientists focus on the institutional barriers.

[20] Nickolas Pappas reminds me that the word "translation" was used in the early Christian church to refer to the translation of, for example, a bishop from one diocese or church to another. This is a fascinating non-linguistic meaning of "translation," but not well known enough to lie behind "translational medicine."

translation, which is typically done by a single linguistic expert sitting at a desk. It is productive to look at why the metaphor of translation seems to resonate with those who use it, and notice whether the metaphor leaves anything unsaid, or is in any important way misleading.

Additional metaphors for the work of translational medicine are also quite common. From the beginning, translational medicine was seen as alleviating the "roadblocks" in the NIH Roadmap for the future (Zerhouni 2003). The "roadblocks" are, fortunately, frequently specified. They include regulatory difficulties and difficulties with developing productive interdisciplinary collaborations. Going from "the bench to the bedside and back" (perhaps itself a spatial metaphor) is intended to describe the scientific journey (indeed, another metaphor) from basic research to clinical applications and back again. The details of how to accomplish this journey are mostly unspecified.

Translational medicine is also described analogically as an "enzyme" for the attempt to "catalyze" the medical applications of basic biomedical research (Nussenblatt et al. 2010, Wang & Marincola 2012), suggesting that medical applications of basic research are being developed too slowly and can be speeded up. Recall that the metaphor of a catalyst was also used for describing the function of NIH Consensus Development Conferences (see Perry 1988). Impatience fuels more than one methodological initiative!

Metaphoric use of language is not inherently objectionable, although it is often unclear or imprecise. The use of metaphor is particularly common in creative work and in pedagogy; both are contexts in which people are developing concepts that are novel to them.[21] (Recall the heavy use of metaphor and analogy for the development of consensus conferences in Chapter 2.) But all metaphors and analogies break down somewhere, and at those points the new conceptual framework that is inspired by the metaphor or analogy has to take some independent steps in order to develop further.

Concretely speaking, the journey from bench to bedside includes thinking about how basic mechanisms can be co-opted for medical purposes, thinking about how in vitro and animal studies should be adapted for human applications, doing initial Phase I studies, observing

---

[21] See, for example, the work of cognitive scientists such as Margaret Boden and Paul Thagard, who write about the importance of metaphor and analogy in creative work.

the results and perhaps experimenting informally with the technology (from bedside to bench) to get better results, and designing further Phase I studies (Marincola 2003). Translational medicine also plays a role in Phase II and Phase III studies by providing rationales for dosages, placebos and/or controls, and protocols (Lehmann et al. 2003).

The skills required for translational research include some understanding of the basic physiological mechanisms and the ability to tinker with them, knowledge of the potential of drugs and technologies in development, clinical skills, creativity, and the ability to negotiate the ethical and legal regulatory requirements. A team approach is a good way to assemble a critical mass of these skills, and the translational medicine literature spends a good deal of time discussing how to encourage such collaborations.

Perhaps the main purpose of the use of "translation" as a metaphor is to suggest that the task of moving from basic science to medical applications is *doable*, at least more or less, just as some sort of linguistic translation between languages is always possible, however imperfect. It is an *aspirational* use of the term "translation." The metaphor is not intended to be interpreted too literally or too deeply; translational medicine is nothing like translational linguistic work done by individual translators who use a dictionary rather than experiments or clinical trials. It is primarily intended to inspire confidence that applied research will eventually be successful. Much is said in the literature on translational medicine about what kind of work is needed to move from basic research to clinical applications. I will look at this next.

## 7.6  Specific Recommendations in Translational Medicine

The metaphor of translation is important, but fortunately there is more to translational medicine than this metaphor. The translational medicine literature is full of suggestions about how to facilitate translational work in general. The recommendations are, with one important exception, not particularly technical, in that formal techniques, precision apparatus, or even virtual collaborations do not feature—at least not yet.[22] I see the

---

[22] The exception is Martin Wehling's proposal, discussed in Section 7.7.

recommendations as falling into two categories: recommendations for improving research infrastructure, and recommendations for broadening the goals of inquiry.

### 7.6.1 Infrastructure Recommendations

What translational medicine has called for so far are buildings, resources, and programs to encourage basic researchers and clinicians to work together face to face, for their students to get some necessary interdisciplinary training, and for collaborations with industry to be facilitated. That is, it has called for *infrastructure* that supports collaborative, interdisciplinary, and entrepreneurial work. There is a good deal of discussion of obstacles to translational work and how they might be alleviated. The obstacles include rigid disciplinary structures, overspecialized training of new PhDs, funding difficulties, market pressures, legal constraints, and ethical requirements. Marincola, in particular, has suggested that a much less hierarchical and more improvisational system is appropriate for translational research—he cites favorably the "Adhocracy" model of Henry Mintzberg (1981).

Thus translational medicine advocates are assessing social epistemological structures for their conduciveness to producing clinical innovation. They are doing applied social epistemology—re-engineering the scientific enterprise-perhaps for the first time since the birth of the "Big Science" era in the post–World War II years.[23] The suggestions that I have seen are not specific to the translational medicine enterprise, in that they have often been proposed for all sorts of interdisciplinary challenges: they include suggestions to bring the workspaces of different researchers into close physical proximity, suggestions to increase interdisciplinary training and collaboration, suggestions to decrease the hierarchical structure of academe, and suggestions to improve incentives and decrease disincentives for successful innovation. The suggestions are, in my assessment, somewhat traditional in that they emphasize the importance of the daily face-to-face environment despite the availability of new technologies for virtual collaboration and opportunities for increased global travel.[24]

---

[23] "Big Science" includes the Human Genome Project.
[24] Of course, it may be that face-to-face collaboration is the best way for humans to collaborate. But we do not know that yet.

## 7.6.2 *The Goals of Inquiry*

Translational medicine advocates have also called for the recognition that important results of inquiry—results worth communicating in the form of publications—may not be in either of the standard forms associated, respectively, with basic science and evidence-based medicine. The standard form for basic science is evidence for or against a hypothesis about basic mechanisms. For evidence-based medicine the standard form is evidence for or against the effectiveness of an intervention in a Phase III clinical trial. That is, translational medicine advocates plead for recognizing—indeed rewarding—*earlier stages* of research, including its *messy* and sometimes *provisional* or *disappointing* moments. Two kinds of appeal are made: first that translational research not be held to the standards of basic research, and second that translational research not be held to the standards of Phase III clinical trials (that is, the clinical trials that produce the important evidence for evidenced-based clinical medicine). I will talk about them in turn.

Basic research makes use of in vitro or in vivo experiments to confirm or disconfirm hypotheses about basic mechanisms. The methodology is familiar to scientists and philosophers of science: it is called "hypothesis-driven research." Typically, hypotheses are used to make predictions about the outcome of laboratory experiments, and the experiments serve as a test for the hypotheses, producing confirming, disconfirming, or inconclusive evidence.

Translational work seeks to do in some ways more and in some ways less than basic research: more, in that some *steps toward successful intervention* are necessary (one can understand a process from basic research without being able to intervene in it, as Hacking (1983) argued long ago); less, in that no particular hypotheses need be confirmed or disconfirmed, and the current understanding of what is going on need not be correct or even known. Typical titles of publications in translational medicine are "Glutamate carboxypeptidase activity in human skin biopsies as a pharmacodynamic marker for clinical studies"; "Angiogenic properties of aged adipose derived mesenchymal step cells after hypoxic conditioning"; "Identification and manipulation of tumor associated macrophages in human cancers" (these titles are all taken from 2011 publications in the *Journal of Translational Medicine*). Such publications provide useful findings for those working on the development of clinical interventions, but are not yet clinical interventions, nor are they basic research, because their value is in pointing the way to how a clinical intervention might work or might be tested. Judging from the range of the titles, "usefulness" does not have explicit standards. (I would argue that it does not need them and that usefulness can be determined in context.)

Translational work also seeks to do in some ways more and in some ways less than evidence-based medicine. As discussed in Chapter 5, evidence-based medicine has limitations as a methodology for producing medical knowledge. In particular, it is a method that devalues mechanistic reasoning, in vitro and animal studies, and indeed everything except for high-quality clinical trials. But the high-quality clinical trials that characterize evidence-based medicine are in fact the final stage of the research process, which begins with mechanistic reasoning and laboratory trial and error and continues with the design of the high-quality clinical trial. When evidence-based medicine is highly valued and rewarded (with grants, publications, and so forth), researchers are encouraged to move to Phase III clinical trials as soon as possible, and may not have the time—or the tolerance for risk—for the problem solving, careful observation, tinkering, and thought that go into the translational work of developing a genuinely new technology.[25] A complete epistemology of medicine needs both translational medicine and evidence-based medicine, as well as a foundation in basic science.

In "Broadening the view of evidence-based medicine," Don Berwick (2005: 315–16)—a well-known champion of evidence-based medicine— argues that we have "overshot the mark" with evidence-based medicine and need to focus more on what he calls "pragmatic science." "Pragmatic science" includes methods such as taking advantage of local knowledge, using open-ended measures of assessment, and "using small samples and short experimental cycles to learn quickly" (2005: 315–16). Berwick did not use the term "translational medicine"—perhaps the term was not yet in wide enough usage—but it is clear from the context that "pragmatic science" involves the same kind of observations and trial and error

---

[25] The fact that much clinical research is supported by pharmaceutical companies who want to see rapid returns on their investments adds to the pressure to skip the difficult steps for real innovation.

experimentation as does translational science. In some[26] of the early papers on translational medicine, we hear similar methodological points about the use of early trials.

Francesco Marincola, the editor of the *Journal of Translational Medicine*, wrote in the first issue that "Translational research should be regarded as a two way road: Bench to Bedside and Bedside to Bench." The "Bedside to Bench" direction requires strong clinical, especially observational, skills. Marincola also notes that "The heart of translational research resides in Phase I trials." But he laments that not enough is learned from Phase I trials because "there is no room in prestige journals for negative results" and because the data in Phase I studies are often "of compromised quality and not of the pristine quality achievable in the pre-clinical setting." He complains that scientists are not inclined to learn "why things did not work" (Marincola 2003: 1). So Marincola is complaining that instead of trying to figure out what went wrong and why, researchers move on to testing the next drug or therapeutic technology. He also argues that descriptive research—especially of unexpected clinical observations—is valuable in itself (Marincola 2007).

Notice that Marincola emphasizes that "The heart of translational research resides in Phase I trials" (2003: 1). Let me ask the question, where is the heart of evidence-based medicine? The answer is: clearly in Phase III clinical trials, which are trials of effectiveness. Marincola is implicitly acknowledging that translational research is not evidence-based medicine. He is more explicitly focusing on the fact that it is not basic research, which he understands as hypothesis-driven mechanistic inquiries into basic mechanisms. He faults basic research for spending too much time on in vitro experiments and animal models instead of on what are needed: in vivo human studies.

Thus, the advocates of translational medicine argue that it should not be held to the standards of either basic science or evidence-based medicine. They hold translational medicine to other standards such as usefulness in clinical intervention, although these standards are generally not made explicit (and I would argue do not need to be). It is not required that publishable results in translational medicine confirm hypotheses

---

[26] Not all. Some of the translational medicine literature emphasizes other reasons for failure to translate basic science into medical treatments, such as "complexity," lack of researchers with the necessary expertise, lack of cooperation between researchers, and insufficient interdisciplinary work.

in basic science, or that they provide high-quality evidence of a finished therapeutic product. To be successful, translational medicine requires more intellectual engagement with mechanisms than evidence-based medicine, and more ability to intervene with natural processes than basic science. I suggest that the new term "translational medicine" gives Donald Berwick's "pragmatic science" the scientific respectability it needs in an age of evidence-based medicine. The term "pragmatic science" might also have worked; the point is that having *some* new term was helpful rhetorically.

Translational medicine is also a response to the disappointments with basic science, specifically the Human Genome Project of the 1990s. The Human Genome Project was accompanied by optimism that sequencing the human genome would bring rapid medical advances in the treatment of genetic diseases, including cancer. The Human Genome Project was described as a "roadmap to health" (Wexler 1992). The optimism, whether justified or not, played a role in raising money in support of the Project. However, there were many disappointments in the early years of genetic therapies. In vitro trials often failed in vivo (Lindee & Mueller 2011). The 1999 death of Jesse Gelsinger in a gene therapy Phase I trial was the final blow to early work in gene therapy. Translational medicine offers hope for the future: it frames the Human Genome Project as basic science—vital research, but not enough for clinical innovation. A new term—"translational medicine" is helpful for describing the trial and error work that comes next and for justifying requests for more funding. Pardridge (2003) argued that the failure to develop therapies from the discovery of the cystic fibrosis gene was due to the underdevelopment of the translational sciences. This explains past failures at the same time as it offers hope for the future.

Theory confirmation in basic science and evidence reviews in evidence-based medicine each offer (different) models of precision (or formality of methods). In contrast, translational medicine offers no such precision. It uses skills such as trial and error reasoning and experimentation to get its results. As suggested in Chapter 6, there are resonances with Hans Reichenbach's (1938) distinction between contexts of discovery and contexts of justification. Translational medicine occupies the place of the context of discovery; theory confirmation in basic science and evidence reviews in evidence-based medicine occupy the place of the context of justification.

## 7.7   Martin Wheling's Critique of Current Work in Translational Medicine

Martin Wehling (2006a, 2006b, 2009, 2011) together with his post-doctoral researcher Alexandra Wendler (Wendler & Wehling 2010, 2012) offer some rare and thought-provoking criticisms of current work in translational medicine. Wehling is the only researcher that I have found who goes so far as to state that most of the discourse about translational medicine is "just wishful thinking" (2008). He does not think that it is sufficient to note the gap that translational work is supposed to fill, or even to gesture in the direction of tinkering and careful observation (as, for example, Marincola does). He wants us to discover what he calls "scientific" methods—by which he means specific, explicit, and objective algorithms[27]—for going from bench to bedside. His central claim is that success "can only be achieved if the translational processes are scientifically backed up by robust methods, some of which still need to be developed" (2008) Indeed, he decries the lack of robust and systematic approaches in translational research and asserts the need for "standardized, reproducible and universal" rules of translation (2006a), making for a seamless transition between basic and clinical research. In a series of recent articles, some with Alexandra Wendler (Wehling 2009, Wendler & Wehling 2010, 2012), they begin to develop the "scientific backbone" that Wehling thinks translational medicine needs. Wehling thinks that knowledge of the appropriate biomarkers for drug function, and assessment of their "predictive values," are key (2006b). By "predictive value" he means the likelihood that in vitro (or animal) testing of the action of a drug on a biomarker will be predictive of in vivo testing in humans. The idea is that clinical research should start with drugs having action on the biomarkers with the highest predictive value. In other words, Wehling believes that translational research can be carried out with more rigor, and that this rigor will help us select those drugs most likely to succeed in subsequent testing. He thinks it will be possible to usefully measure therapeutic potential quantitatively. In Wendler and Wehling (2012) a "translatability score" is developed based on the predictive value of biomarkers found in in vitro and animal experiments.

---

[27] This is not my definition of "scientific." As I have been arguing throughout this book, science also has messy and tacit elements.

At this time, Wehling's proposal is in a preliminary stage and it is premature to evaluate it definitively. It is worth pointing out, however, that it assumes that biomarkers have the same predictive values for different therapies which affect the same biomarkers. This assumption will be tested empirically when attempts are made to determine the predictive values for each biomarker. It also assumes that we know enough to meaningfully estimate the predictive values of biomarkers and to reach consensus on such estimates. Usually, however, we have little such knowledge in new areas of research.

Interestingly, Wehling recommends that the scoring of biomarkers be achieved through the consensus of experts (Wehling 2009: 544). Here is yet another case where expert consensus is appealed to for epistemological foundations.

Translational medicine is difficult in large part because of the *complexity* and *variety* of mechanisms in vivo. Interventions that work in vitro and in laboratory animal models often do not work in human systems, which may share some basic mechanisms but not all relevant mechanisms. If we had a complete model of human biological processes in all their complexity, we might be able to predict pharmaceutical success and turn translational medicine into the kind of rational process of discovery that Wehling would like. Since we do not, it is more of a process of trial and error. Wehling is putting the cart before the horse, assuming that we can have adequate knowledge of a new intervention to usefully predict its success. In my view, trial and error is typical of the discovery phase of scientific inquiry, and, while we may find some worthwhile heuristics (such as looking at affected biomarkers), we are unlikely to find the kind of "robust methods" that Wehling is seeking.

## 7.8  Illustrative Example: Successes and Failures of Translational Medicine in Cystic Fibrosis Research

In Chapter 5, I described recent research on effective treatments for cystic fibrosis. In that chapter, my goal was to show the limitations of evidence-based medicine. The techniques of evidence-based medicine were vital for assembling evidence for the successful treatments, which have prolonged the lives of cystic fibrosis patients from infancy to early middle age. Until very recently, treatments have been empiric and non-specific to cystic fibrosis in that they have addressed symptoms such as congestion with already known techniques such as chest percussion. The gene underlying cystic fibrosis, CFTR, was identified in 1989 and genetic therapies were attempted in the 1990s, but they did not work in vivo despite high hopes and promising in vitro studies. In twenty-first century vocabulary, this is described as translational failure. Together with other translational failures in genetic technology (and especially with the Jesse Gelsinger case, in which a healthy experimental subject died), it led to a period of discouragement about genetic therapies.

In the past decade, the term "genetic therapy" has been broadened to include genomic and proteomic therapies. All these "omic" therapies work at the proximal end of the causal chain, tinkering with the basic molecular mechanisms responsible for the disorder. New strategies—strategies other than fixing the basic gene defect—have been attempted. For cystic fibrosis, attempts have focused on fixing the defective CFTR protein (not the gene that codes the protein, which proved resistant to replacement in early trials) and on enhancing chloride transport (the functional abnormality produced by the defective CFTR protein).

In Chapter 5 I described several attempts to intervene with genetic, genomic, and other molecular mechanisms underlying cystic fibrosis. All the attempts were supported by plausible mechanistic reasoning and either in vitro or in vivo studies. These attempts are all examples of "translational medicine" because they aim to translate basic mechanistic understanding into effective interventions. Some seem to work: Kalydeco (VX770) seems to have safety and effectiveness for the G551D mutation, and Kalydeco in combination with Lumacaftor (VX809) has safety and effectiveness for the common deltaF508 mutation. Kalydeco and Lumacaftor are now FDA-approved treatments for cystic fibrosis. Their safety and effectiveness was suggested in Phase I and II trials, and supported in Phase III trials. Other attempts to intervene have been less successful: PLASmin and Denufosol failed to show effectiveness in Phase III trials and Ataluren's results have been mixed. (I explained the intended molecular mechanisms for these drugs in Section 5.7.)

Was it possible to predict in advance which of these drugs were likely to fail and which were likely to succeed? At this time, no. As discussed in Section 7.8, Martin Wehling (2009) hopes that it will be possible in the future. In Section 5.7 I expressed some skepticism about the possibility of predicting success in advance, arguing that unless we have a complete theory of everything (which we do not) we will continue to be surprised by the results of clinical trials. Of course, the more we know about basic and other mechanisms, and the more we know about comparative physiology (of laboratory animals and humans) the more likely we are to make accurate predictions and avoid drug failure by focusing on those interventions with the greatest probability of success. But in this context of discovery, we are necessarily working without full knowledge of all relevant mechanisms. (We may have full knowledge of what we think of as *basic* mechanisms, but this is not enough when working in vivo or when "translating" animal to human studies.) I think that some unpredictable failure is unavoidable.

Reasoning about therapeutic interventions in cystic fibrosis has the kind of trial and error character described by Berwick and by Marincola in Section 7.6.2, rather than the kind of "rational" and systematic process of prediction that Martin Wehling has in mind. Researchers spend time trying to figure out why an intervention did not work. Drugs are sometimes tried again at different dosages, or with the addition of "promoters" or "enhancers" designed to address hypothesized interferences with the desired mechanisms of action. A good deal of thought and tinkering goes into designing the next Phase I and Phase II clinical trials.

## 7.9 On T2

T2 is the challenge of moving effective new medical interventions into regular clinical practice ("bedside to community"). It is a challenge that was recognized as early as the 1970s, when it was hoped that the NIH Consensus Development Program would facilitate the dissemination of new medical research. At that time, the challenge was thought to be simply *epistemic*; clinicians were hypothesized as ignorant of or uncertain about newly developed health care interventions. As discussed in Chapter 2, the hope was that a well-publicized expert consensus would "close the gap" between current knowledge and practice. But from the beginning (Jacoby 1983) dissemination of knowledge was poor, and even where there was dissemination, incorporation of the knowledge into regular practice was lacking (Kosecoff et al. 1987). Efforts to disseminate the results of consensus conferences were increased through direct mailing to physicians, publication in prominent journals, and, more recently, a comprehensive website. But this made only a little difference (Ferguson 1993). Efforts to disseminate the results of evidence-based medicine have been similarly disappointing. Moving effective new medical interventions into regular clinical practice is a continuing challenge. New approaches—such as guidelines, incentives, outcomes research, social networks, and so forth are being devised (see, for example, Gawande 2013), since it has become clear that passive knowledge dissemination is not sufficient.

In Chapter 3 it was noted that dissemination is achieved somewhat better in countries that have more centralized organization of medical delivery, such as Canada and France. It is likely that recommendations for improvement of dissemination will be different for different countries and different contexts.

T1 and T2 have little in common with each other. They are different kinds of challenges. Both are called "translational" challenges because the metaphor of translation is deemed good enough (one might say vague enough) to describe the process of moving from one kind of knowledge to another. T2 is sometimes broken down into stages, for example from evidence-based medicine results to evidence-based practice guidelines, from guidelines to regular health practices and Phase IV clinical trials, and from regular practices to outcomes research (Khoury et al. 2007). I submit that just as T1 became timely after a decade of disappointing clinical results in genetic technology, T2 became timely after three decades of limited success at diffusion of medical innovation through consensus conferences and clinical guidelines. Again, we have a new name and a new strategy for addressing an old problem: trying to improve regular clinical practice by implementing new results of research. T2 is trying to do what consensus conferences tried to do, with the advantage of some new ideas about how to bring about practice change.

## 7.10  Conclusions

Translational medicine is a new term for endeavors that are as old as scientific medicine: the typically difficult and messy work of going from basic research to medical applications (T1), and the slow work of moving successful new interventions into regular clinical practice (T2). Having a new term, "translational medicine," is helpful because the necessity for this work was obscured at the end of the twentieth century for three reasons. First, there were wildly optimistic expectations of the basic science work of the Human Genome Project; second, there was lack of recognition of the incompleteness of evidence-based medicine methodology; and third, there was ignorance about what it takes to disseminate new practices. It is helpful to have the new term, "translational medicine," because it restores hope in future clinical discovery, acknowledges the importance of non-formal methods in science, and validates the need to devise social epistemological interventions that specifically support translational work, both T1 and T2. The methodology of T1 is currently and probably unavoidably rather inexact, precisely because it is the context of discovery. In addition, failure is an expected part of its work of discovery of new and successful interventions. Good methodology for T2, on the other hand, may become standard as we learn more about the processes of diffusion of medical interventions. Even so, it is likely to be contextualized to particular national, cultural, and professional conditions.

# 8

# On Narrative Medicine

## 8.1 Introduction

Narrative medicine is the most prominent recent development in the medical humanities.[1] Its central claim is that attention to narrative—in the form of the patient's story, the physician's story, or a story co-constructed by patient and physician—is essential for patient care. This is a stronger claim than the frequent exhortation to "listen to the patient," not only because it includes physician narratives but also because there is explicit attention to the *narrative form* of the stories as well as to their content. Narrative form contains information that is relevant to treating the individual patient, and attention to narrative form contributes positively to the physician–patient relationship. The claim is that "narrative competence"—the ability to construct as well as understand (deconstruct) stories—is part of clinical expertise.

Other recent developments in the medical humanities include hermeneutic, phenomenological, empathic, and "embodied experience" approaches.[2] There is much overlap between all of these methods. For example, a narrative used in narrative medicine can report an embodied experience and in this way be an example of a phenomenological account. Or, for example, a physician with hermeneutic goals is likelier to be empathic. To keep the discussion manageable, my emphasis in this chapter will be on narrative medicine.

---

[1] This chapter develops some of some of the themes of Solomon (2008).

[2] Hermeneutics and phenomenology are somewhat overlapping philosophical traditions that emerged in the late nineteenth and early twentieth century and continue to the present. Recent work on empathy has roots in the hermeneutics tradition and recent work on embodied experience has roots in the phenomenological tradition.

Narrative medicine is a locus of typical claims about the necessity for an art as well as a science of medicine. Narrative medicine is the "art," and "the biomedical model" and/or evidence-based medicine are cast as the science. As discussed in Chapter 1, such claims are of limited usefulness, first because they encourage a dichotomous classification of methods which obscures an underlying, richer, pluralism, and second because the claims tend to go together with traditional but misleading views about science, such as that science eschews narrative in favor of causal conditions and logic. In fact, narrative is quite common in science. For example, late nineteenth-century theories of evolutionary change had a variety of narratives of progress (or, in the case of Darwin, the lack of progress). Cosmological accounts tell narratives of the history of the universe. Harry Hess's "seafloor spreading hypothesis," which was the beginning of plate tectonics, was in a narrative form.[3] To be sure, more formal methods, such as Mill's methods, may be used to justify causal hypotheses, and more precise and general forms of discourse, such as numerical equations describing scientific laws, may be used to express mature theories. But narratives are a common form in which we discover, state, and reason about causal connections. The usual requirements of narrative coherence and inclusion of significant facts put some constraints on the stories that are told, making a good narrative an intellectual achievement that can pave the way, when appropriate, for more precise and more general causal hypotheses that have tighter constraints. Narrative reasoning, in my view, does not employ a *sui generis* logic[4] wholly different from the logic of science. It makes frequent use of causes as well as temporal order in its accounts and explanations of change.[5]

General narrative competence includes a basic understanding of narrative theory, especially reader-response theory in which meanings emerge through the reader's interpretation of the form and the content of the narrative. The narratively competent reader should be able to detect narrative elements such as setting, plot, frame, tone, and desire and narrative types such as restitution, chaos, and quest (redemption).

---

[3] He called it "geopoetry" in Hess (1962).

[4] Kathryn Montgomery (2006) and Rita Charon (2006a) would disagree with me here. They think that narrative reasoning and scientific reasoning are completely different kinds of reasoning. Some of the ideas in this chapter reflect our different understandings of the nature of science.

[5] Those using narratives in a scientific context often treat reasons as causes.

This requires emotional sensitivity and responsiveness as well as cold (non-affective) cognitive skills. Specific narrative competencies for clinical medicine will be discussed in the Section 8.2.

Rita Charon is perhaps the most well-known proponent of narrative medicine. She makes the most ambitious claims about its relevance to clinical medicine and also makes the most thorough use of tools of literary analysis. Her book *Narrative medicine: honoring the stories of illness* argues persuasively for the idea that "good readers make good doctors" (2006: 113) and for the inclusion of narrative studies in medical education. She is the Director of the new Program in Narrative Medicine at Columbia University which offers a Master of Science degree in the field. Her work is a culmination of several decades of attention to narrative in medicine starting with Howard Brody (2003 [1987]), and continuing with the work of Arthur Kleinman (1988), Kathryn Montgomery (1991, 2006), Arthur Frank (1995), Trisha Greenhalgh (2006), and others. At the same time, there has been focus on narrative in ethics—particularly medical ethics—also starting with the work of Brody (2003 [1987]) and continuing with the work of Hilde Lindemann (1997) and others. More generally, narrative approaches have become important in philosophical accounts of personal identity and in the qualitative social sciences. Narrative medicine is part of this more general academic turn to narrative.

Stories about health and illness have been told since ancient times in all cultures. They are often interesting and memorable and have always played a part in medical pedagogy. The current interest in medical narrative goes beyond this to argue that there is important, medically relevant information in the form as well as the content of illness narratives, and that both listening to and co-constructing narrative is part of the work that the physician does to heal the patient. In narrative medicine, narrative has moved to the center stage of the medical encounter and narrative competence is one of the major skills that go into clinical expertise.

In Chapter 6 I described how "clinical expertise" is needed for applying medical knowledge to particular cases. In that chapter I focused on the importance of the ability to handle biological variability, which is part of clinical expertise. Care needs to be individualized whenever we have reason to think that such biological variability matters. In this chapter I look at other aspects of "caring for the individual patient," specifically those aspects in which narrative medicine claims to be required: taking a good patient history, establishing a therapeutic relationship between

doctor and patient, incorporating patient preferences and patient autonomy, and addressing the patient's suffering. There is now a substantial
literature arguing that evidence-based medicine can and should be
integrated with narrative medicine for complete patient care (see, for
example, Charon et al. 2008, Meza & Passerman 2011).[6] Narrative medicine is partly about handling *psychological* variability: the physician
needs to find out and take into account the preferences and values of
the patient. But it is also about the physician–patient *relationship*, and
how the humanistic aspects of that relationship can address some of the
patient's needs.

## 8.2  Narrative and Narrative Competencies
##      in Medicine

Narrative is story. Telling a narrative is telling a sequence of connected
events. Traditional narratives have a beginning, a middle, and an end;
nontraditional narratives can have a different structure. Charon defines
narrative somewhat traditionally as "stories with a teller, a listener, a time
course, a plot and a point" (2006a: 3).[7] Broadly speaking, narrative does
not require language: narratives can be told in dance, art, and music as
well as orally and in writing. The form of the narrative—voice(s), tense,
tone, and so forth—also conveys information and meaning. Narratives
are about particular events, rather than about events in general, although
several narratives can share the same narrative type in that they can be
the same kind of story. There is no precise definition of narrative or even
of medical narrative. For reasons that I will explore, the field of narrative
medicine, in practice, defines the scope of narrative as widely as possible.
    Narrative medicine includes the telling and analysis of stories of ill-
ness, such as those told by patients and doctors, but often goes further to
include any discourse (or activity) that has narrative elements, whether
or not it qualifies as a story in the colloquial sense. For example, Rita
Charon regards the patient chart as a kind of narrative, noting that its

---

[6] I even found one article that claims to integrate evidence-based medicine, narrative
medicine, and translational medicine (Goyal et al. 2008).

[7] In practice, her definition of narrative is broader and includes, for example, analysis
of the hospital chart as a narrative. Hospital charts do not have a point (unless they all have
the same point).

temporal organization and its passive voice convey meaning.[8] She also analyzes patient monologues (and dialogues with the physician) as narratives, even if other elements such as argumentation, description, and theorization are present and prominent, and even if the monologues (or dialogues with the physician) are so disorganized that a traditional narrative is not present.

I have discerned several levels of narrative competence in the practice of medicine, and in the following sub-sections I will classify them into four types and illustrate them with some examples from the narrative medicine literature. The first two types of competence are narrative medicine in only the broadest sense, where "narrative" covers whatever way the patient chooses to communicate with the physician about their situation. The two types of competence that follow make use of more specific narrative tools. The levels of narrative competence differ in hermeneutic depth, and begin with the most basic. My classification is for the purpose of my own exposition and discussion; other classifications might work equally well for different purposes.

### 8.2.1 First Narrative Competence: Listening and Witnessing

Narrative medicine is not only the exhortation to "listen to the patient," but being attentive to what the patient communicates is a necessary component and a starting point. Listening, and appearing to listen (which is also important), requires general communicative skills and attention to both what the patient says and how the patient says it. Good listening is also a form of *witnessing*, which in turn is an important way of providing relief for suffering.

A typical example, from Rita Charon, is the case of a 36-year-old Dominican man with a chief symptom of back pain. When invited to tell the doctor all that he thinks she should know about his situation, he starts telling a complex story and then "after a few minutes he stops talking and begins to weep. I ask him why he cries. He says 'No one has ever let me do this before'" (2004: 862). This is a touching story, made more real by the details of age, gender, and ethnicity.[9] It shows the therapeutic

---

[8] In Charon's view, the passive voice communicates professional detachment.
[9] I note with a little irony that these narrative details are changed to protect the identity of the patient. Moreover, the same example appears almost verbatim with a 46-year-old Dominican man with shortness of breath and chest pain in Charon (2006a: 177).

benefit to the patient of the physician listening to their story in all its detail and complexity. It bodes well for the development of a positive physician–patient relationship, in which any technical interventions will be used in a context of trust and openness.

A less typical example is Clifton Meador's "The woman who would not talk" (2005: 114–23).[10] In this example, the patient was unwilling to speak, and the physician established communication through mirroring the patient's bodily movements and encouraging her kinesthetic processing. The example is worth noting because it shows an unusual and creative mode of listening, one that borders on empathy (which will be discussed in Section 8.2.2).

Listening is obviously important for gathering information (physical, psychological, and social), and most exhortations to "listen to the patient" have such information gathering in mind. Narrative elements (such as mood, tone, desire) in what the patient communicates, whether verbally or non-verbally, may provide further information. Examples are given in Section 8.2.3. The point of this section is that listening and witnessing—independent of the content of the narrative—can be therapeutic.

Listening is more than sitting quietly while someone else speaks, even though such mere appearance of listening can be of some benefit to the speaker. Listening requires, minimally, attention and the ability to be moved in some way by what the speaker says.

### 8.2.2  Second Narrative Competence: Empathy

The second narrative competence is an intensification of the first; not just the ability to listen to the patient, but the specific ability to understand *at an experiential level* what the patient is going through. This ability is often called empathy, and described as the ability to imagine *what it is like* to be some other person. In particular, empathy includes the ability to experience vicariously (although at less intensity) what the other person is experiencing emotionally. Empathy (in common with sympathy and pity) involves emotional rather than (or in addition to) intellectual engagement. Empathy is, however, distinguished from the experience of

---

[10]  I am grateful to Steve Peitzman for bringing this example to my attention.

emotions such as sympathy and pity, which are generated by second- or third-hand knowledge of the situation of the other person.[11]

Jodi Halpern's *From detached concern to empathy* (2001) is an extended treatment of the role of empathy in clinical expertise. It has a central case study, that of Ms. G., who refuses the dialysis necessary to sustain her life and dies as a result. Halpern was involved in this case during her training in psychiatry (her eventual specialty). Ms. G. suffered from diabetes and kidney failure and had recently undergone a second leg amputation. At the same time, her husband left her. Halpern tried to acknowledge Ms. G.'s despair by starting a sympathetic conversation about Ms. G.'s losses, but Ms. G. became angry and refused further conversation. Clearly this case has haunted Halpern, who feels that somehow she, or other physicians, should have been able to intervene more effectively while still respecting the patient's autonomy. Halpern suggests that the physicians involved in Ms. G.'s care failed to empathize with her and instead reacted from their own, ignorant perspectives of what it would be like to be in Ms. G.'s position. She reports quick judgments, such as that of her internist: "Wouldn't you want to die if you had lost your spouse, your legs, your kidneys, and faced a future of blindness and other medical problems?" (2001: 3). Such judgments generate, at most, emotions of sympathy and pity, which were not helpful to Ms. G. Halpern thinks that if she had taken the time to really empathize with Ms. G., she would have discovered that Ms. G. was going through a grief reaction for which support and understanding (rather than pity and sympathy) would have been helpful. Halpern argues that the grief reaction should not have been taken as a plain statement of Ms. G.'s true wishes and should have been explored before withdrawing dialysis treatment.

According to Halpern, successful empathizing gives the physician access to the patient's emotional reasoning, which she describes as "linking ideas that have affective, sensory and experiential similarities rather than logical similarities (2001: 41). Charon similarly describes narrative as constituting "a logic in its own right" (2006a: 41) that is different

---

[11] This is a set of distinctions that is standard in the literature, and I will work with them, although I am not personally convinced that empathy is always different from sympathy and pity because how we feel about ourselves is connected to how we feel about others.

from the reasoning used in science. Attempts to empathize should avoid
the pitfalls of focusing on either the doctor's own emotional reasoning
or the doctor's logical reasoning. Halpern writes, "My central claim is
that empathy requires experiential, not just theoretical knowing. The
physician experiences emotional shifts while listening to the patient"
(2001: 72).

Although Halpern does not mention it, her use of empathy as a source
of knowledge about patients is similar to the recent "simulation theory
of mind" in philosophy of psychology, as put forward by Vittorio Gallese
and Alvin Goldman (1998) and Goldman (2006). The claim is that in
addition to their explicit theory or theories of the mind, we anticipate
mental states in others by simulating their mental processes ourselves,
and we are often successful because of the similarities of our minds. Some
empirical confirmation of this theory has been obtained in research
on monkeys by Rizzolatti et al. (2006) and in research on humans by
Iacoboni et al. (2005).

I classify empathy as a narrative competence. This is because one of the
purposes of narrative is to try to convey what an experience is like. The
patient's narrative provides the information that makes successful empathizing possible. Otherwise the physician is just guessing at what the
patient is experiencing. Charon insists on careful listening to the patient
as part of the method of achieving empathy. She describes the process of
empathizing as follows:

> By emptying the self and by accepting the patient's perspectives and stance, the
> clinician can allow himself or herself to be *filled* with the patient's own particular
> suffering, thereby getting to glimpse the sufferer's needs and desires, as it were,
> from the inside. (2006a: 134, original emphasis)

Listening and witnessing can be undertaken at different depths. Empathizing is perhaps the deepest kind of listening and witnessing because
it is the attempt to fully experience what the patient is experiencing.[12] This
requires emotional availability from the physician. Metaphor is often
used when explaining empathy, as exemplified by Charon's description
of "emptying the self" so as to understand the experience of the other.
Charon also describes this as a process in which "doctors use the

---

[12]  Halpern and Charon do not talk about whether or not (or how to tell that) empathizing achieves these goals successfully. Can the physician really "empty the self"? Does the
physician really experience what the patient is experiencing?

self as a potent therapeutic instrument" (2006a: 194). Empathizing is engaged emotional work, and very different from the "detached concern" traditionally expected of the health care professional.[13]

Havi Carel (2008) and Kay Toombs (1992) have powerfully argued, through their own illness narratives, that it is important for the physician to understand the experience of the patient, from the patient's perspective, calling this "embodied experience" and invoking phenomenological methods. Their greatest complaint is that clinicians lack empathy, and suggest that because of this lack clinicians do not address the patient's suffering and indeed contribute to it by making insensitive remarks. Carel and Toombs attempt to communicate their embodied experiences of illness and disability. For example, distances can appear greater than usual because of difficulty with mobility, while spaces seem to contract when navigating with a bulky wheelchair. Felt agency[14] is reduced and the world can be experienced as obstructive. Time is experienced more in the present, in the extra attention needed for dealing with mundane tasks. Over time there can be adjustment to a "new normal," and the experienced quality of former abilities may or may not be remembered. (For example, Carel reports dreaming of running, while Toombs cannot remember what it feels like to walk.) In the social world, Toombs and Carel are astonished by the insensitivities of other people, including most members of the medical profession. Toombs also reports her emotions of shame about her disabilities; she thinks that shame is an irrational emotion. Both clearly feel highly vulnerable.

The use of empathy in narrative medicine also has similarities with the methods of the nineteenth-century *Verstehen* tradition in philosophy. The *Verstehen* tradition is the ancestor of both hermeneutic and phenomenological methods, and distinguishes the human sciences from the natural sciences by their need for empathic skills. Toombs and Carel explicitly identify their approaches with the phenomenological tradition. The psychiatrist Eric Falkum (2008) connects his similar approach with the hermeneutic tradition. Fredrik Svenaeus (Svenaeus 2000) connects

---

[13] At this point a criticism is often made that narrative medicine requires too much of the contemporary clinician, who has little time for each patient. Both Charon and Halpern address this criticism, arguing that full empathizing is not necessary in all clinical encounters, and that when it is necessary it can save time overall by focusing attention on the important issues.

[14] Felt agency is the feeling of being in control.

his "philosophy of medical practice" with both the hermeneutical and phenomenological traditions.

Empathy, which requires emotional engagement and a willingness to see a subjective point of view, does not look like traditional scientific method. From the beginning of the *Verstehen* tradition through current work in hermeneutics and phenomenology, empathy is contrasted with the supposedly more "objective" methods used in the natural sciences. I will conclude this section by arguing that this apparent contrast does not survive in contemporary philosophy of science.

No methods used in the natural sciences are perfectly "objective"— by which is usually meant value free, neutral, unemotional, and conducive to truth.[15] There is also no single, general scientific method that applies always and in all domains. Empathy can be regarded, straightforwardly, as one of the fallible approaches that we use to understand and predict. It is not an explicitly stated method—but no scientific methods are explicitly stated in full. It is particularly useful with regard to human behavior, and may also be helpful for grasping the behavior of higher animals.[16] Beyond that, it may be helpful in areas not involving minds at all. Evelyn Fox Keller's (1983) study of Barbara McClintock suggests that McClintock's empathy with chromosomes viewed through the microscope played an important role in her process of discovery of genetic transposition. Arthur Fine reports observing physicists and chemists reasoning about particles and molecules by pretending to "think like" them.[17]

In addition, empathy is not the only method we have for understanding and predicting human behavior. These days we have a variety of tools, including psychoanalytic theories, behavioral theories, cognitive theories, behavioral genetic theories, and neural theories. Empathy does not characterize the whole methodological scope of the human sciences; *Verstehen* is just one of several methods.

Skeptical questions about what is accomplished during attempts to empathize are philosophically familiar. One way of expressing such

[15] This claim depends on the work of the Kuhnian revolution and on the work of feminist science critics.

[16] Current "canine philosophy" is that if you want to train a dog, then you have to think like a dog.

[17] Arthur Fine made this remark on November 9, 2007 after I presented some of this work at a talk at the University of Washington, Seattle.

skepticism is to invoke Thomas Nagel's "What is it like to be a bat?" paper (1974), which is well known to philosophers. In this paper, Nagel argues that there is no scientific evidence relevant to the question of what it is like to be a bat. Bat experience is inaccessible to us, even if we know much about bat behavior and bat neurons. This argument can be extended to what it is like to be a particular human, but with a crucial difference. Humans can describe what it is like to be themselves. Insofar as these descriptions can be communicated between people, they can be used to test attempts to empathize. It is possible to discover that we have failed to empathize. For example, we can be mistaken about what a friend who is in trouble wants to hear from us. When we make efforts to test our attempts to empathize (by asking, "Is this how you feel?" or "Is this what you want?") we often discover that our attempts are flawed when our friend corrects our suggestions ("I'm not only sad, but I'm angry," "I'd prefer it if you didn't try to solve my problems for me.") Should these discoveries be used as a basis for improving empathic understanding or for drawing general skeptical conclusions about attempts to empathize? That will depend on the extent of our failure to empathize and our ability to correct such failures.

Why is empathy part of the therapeutic encounter? Empathizing is one (fallible) way of identifying the needs of the patient as well as—or even better than—the patient himself or herself can articulate. The physician is then able to devise a suitable therapy. Empathizing is also a "deep" form of listening in which the speaker is joined in their experience. It is a form of intimacy that can reduce the suffering of the patient by making the suffering a less lonely experience.

### 8.2.3  Third Narrative Competence: Narrative Detective Work

Rita Charon has a number of examples in which she uses her knowledge of narrative form to discover information that is relevant to both diagnosis and treatment. Close reading, with attention to frame, form, time, plot, and desire, yield information about temporality, singularity (particularity), causality/contingency, intersubjectivity (relationships between people), and ethicality. I call this "narrative detective work," and it is taught in literature classes. A typical example is Charon's use of a patient's narrative of a family history of pancreatic cancer to diagnose not pancreatic cancer, but a minor abdominal problem and suicidal wishes. She diagnoses these because she detects in the patient's narrative a surprising certainty about his self-diagnosis and a surprising readiness to die. The "desire" in the narrative contained important information.

The narrative detective work is not always done on the patient's narrative; sometimes it is the physician's narrative that yields important clues. One of Charon's central examples is the case of Luz, a young patient with mild headaches, who urgently requests Charon's signature on a disability form at an inconvenient time. Charon is irritated both by the inconvenient request and by Luz's attempt to use her headaches as a justification for disability payments. She finds herself constructing a narrative to fill in the blanks in her knowledge of the case. Charon's fiction is that Luz is an aspiring fashion model who wants disability payments to tide her over until her career takes off. This story is actually wildly off the mark, but Charon's investment in thinking about Luz's case pays off the next time she sees Luz, when she is primed to ask Luz more fully about her situation. It turns out that Luz was escaping sexual abuse at home and was trying to set up a safe place where she could also bring her younger sisters. Charon claims that her incorrect narrative functioned as "a tool with which to get to the truth" (2006a: 6) and was stimulated by the wordless communication of urgency from Luz to Charon. This is a case where the details of the narrative matter only in so far as they led the physician to seek further information; diagnosis is not available with narrative analysis alone. Such a case is intended to illustrate the benefits of a "parallel chart" to be written by the physician in narrative form alongside the regular hospital chart, for the purpose of physicians' critical reflection on their own practice.

Narrative detective work makes a bold claim: attention to the form as well as the content of narrative is important for diagnosis and treatment. Narrative detective work can yield surprising and satisfying diagnoses and resolutions of clinical problems. However, most of the examples that Charon and others give of "narrative medicine" are not of medical mysteries solved by discovering clues in narrative structure. Far more are about listening, empathizing, and expressing meaning without the apparatus of narrative theory.

### 8.2.4 Fourth Narrative Competence: Making Meaning

In his classic book *The nature of suffering and the goals of medicine* (1991), Eric Cassell argued that it is the responsibility of the physician to address the emotional and existential suffering of the patient as well as his or her physical suffering. Many others—for example, Howard Brody, Havi Carel, Rita Charon, Arthur Frank, Brian Hurwitz, Arthur Kleinman, Kathryn Montgomery, Edmund Pellegrino, and David Thomasma—have argued similarly that medicine is responsible for healing the patient and not only for curing or palliating the disease.[18]

Listening to the patient and empathizing with the patient both address the suffering of the patient, and have already been discussed. A further way of addressing suffering is by attempting to make sense of it. "Making sense" goes further than "understanding the patient" in that it helps the patient get a *better* (more satisfying or more empowering or even more true) understanding. Telling a disease narrative is an important and frequent way in which patients and physicians make sense of suffering. "Making sense" of suffering is accomplished by a variety of kinds of narrative: Arthur Frank has distinguished restitution, chaos, and redemption narratives. Redemption narratives are perhaps the most frequent; they represent serious illness as a fundamental disruption to the life narrative, which is resumed only after struggle and with profound change to the life narrative. One example is the classic breast cancer narrative in which the first reaction is grief ("first you cry"[19]) and confusion, the second stage is determination to undergo grueling treatment, often finding strength in purposes such as living long enough to raise children, and the final stage is emergence with a transformed appreciation of the value of life. The task of constructing or co-constructing a disease narrative makes use of creative narrative competence.

Howard Brody, who was one of the earliest to realize the importance of illness narratives, quotes Oliver Sacks (1985: 12) on the topic of narratives and personal identity in his 1987 book *Stories of sickness* (Brody 2003 [1987]: 2):

Each of us *is* a biography, a story. Each of us *is* a singular narrative, which is constructed continually and unconsciously by, through, and in us—through our perceptions, our feelings, our thoughts, our actions; and, not least, through our discourse, our spoken narrations. Biologically, physiologically, we are not so different from each other; historically, as narratives, we are each of us unique.

---

[18] This claim is of course subject to challenge, and will be challenged in Section 8.5.

[19] The book *First, you cry* by Betty Rollin (1976) was one of the earliest and most influential examples of this kind of narrative. Barbara Ehrenreich (2001) and Judy Segal (2007) have commented critically on this narrative, and I will introduce their ideas in Section 8.4.2.

It is common to find personal identity (the self) identified with personal narrative. We find such claims not only in Brody's book, but also centrally in the work of Rita Charon and others. It seems so obvious to those who work on narrative medicine that they do not bother even to argue that narrative constitutes the self; at most they cite philosophical, psychological, social scientific, or literary sources for the view. For example, Alasdair MacIntyre (1981) and Marya Schechtman (1996) are common philosophical sources. Completing a narrative then comes to stand for making the self whole again.

The claim that one's narrative is one's self invariably goes together with a view that both selves and stories are unique, or, as Rita Charon puts it, "singular." She writes, "I am stunned at how singular—how absolutely unique—are these self/body narratings" (2006a: 188). It is claimed that every story is different, and that every person is different. Persons are apparently valued as individuals in large part because their differences from one another are cherished. The differences are equated with the uniqueness of persons. Would we value people less as individuals if they were more similar to one another? I will look at the concept of "singularity" critically, in Section 8.4.2.

If the narrative is the self, sharing one's narrative(s) is sharing the self, a deeply intimate act. When the physician participates in the construction of patient narratives, he or she needs to try to empathize with the patient, which requires the physician's emotional engagement. Emotional engagement uses the self, making engagement of the physician doubly intimate in that it requires both empathy and the sensitive construction of narrative. Charon believes that "the self is the caregiver's most powerful therapeutic instrument" (2006a: 7).

Making meaning contributes to intimacy. Intimacy in turn produces affiliation (and vice versa). A strong physician–patient relationship is often essential to patient care, since it can contribute to finding a diagnosis, choice of treatment, compliance and effectiveness of treatment. Charon goes so far as to say that "the body will not bend to ministrations from someone who cannot recognize the self within it" (2006a: 182).[20]

Because "making meaning" is seen as part of patient care, and narratives are seen as carriers of meaning that constitute selves, people with

[20] Obviously this is overstated; physicians lacking in interpersonal skills sometimes get excellent results.

illness are readily conceptualized as in need of "narrative repair." Arthur Frank (1995) describes patients as "wounded storytellers" whose narrative voices (as well as whose bodies) are damaged by illness. He writes that "Stories have to *repair* the damage that illness has done to the ill person's sense of where she is in life and where she may be going" (1995: 53)

Thus the narrative competencies required of physicians include listening to and helping with the construction of narrative. The goal is a sense of coherence that comes with having narratives that work.[21]

"Narrative repair" has similarities with the psychotherapeutic modality of "narrative therapy," in which the psychotherapist assists the patient in developing better narratives about their life (see, for example White & Epston 1990). I will say more about this in Section 8.4.6.

## 8.3   The Ethos of Narrative Medicine

Narrative medicine has an "ethos," by which I mean that it has a moral and political character. Its ethos is similar to that of the traditional healer–patient relationship, a one-to-one relationship marked by trust and intimacy. I use the word "healer" to mark the cross-cultural and historical importance of such dyadic healing relationships. In a healing relationship, psychological[22] as well as physical needs are addressed. Narrative medicine, in common with humanistic medicine in general, and with many kinds of alternative and complementary medicine, argues that something like the traditional healer–patient relationship is essential for taking care of patients, and that the "biomedical model" (or "scientific medicine") does not pay enough attention to this.

The ethos of narrative medicine is more democratic—more egalitarian—than that of some traditional healer–patient relationships. This is not surprising, since narrative medicine developed after the Nuremberg Code and after Tuskegee and along with the patients' rights movement. In the place of Osler's "detached concern" is the empathic presence of the physician who allows him or herself to be moved by the patient's narrative. The ideas of Martin Buber and Emmanuel Levinas on

---

[21] Much is skimmed over in the phrase "narratives that work," particularly the distinctions between narratives that are useful, narratives that are coherent, narratives that are true, narratives that are empowering, and narratives that are preferred.

[22] "Psychological" is construed very broadly, to include spiritual and social aspects of experience.

the importance of the I–Thou relationship and the face-to-face encounter are often invoked. Charon's work treats the physician–patient relationship as a relationship of equals as well as an intimate relationship.

Probably the most common complaint that patients make of contemporary health care is that it is impersonal. Havi Carel (2008), for example, reports that that she missed hearing even basic sympathetic acknowledgements such as "I am sorry to hear about your illness." Rita Charon and others focus not only on basic human respect but on the particularity of each clinical situation, deploring "cookbook medicine" as not sufficiently nuanced for the care of individual patients. Treatment of individual patients takes time, caring, and intimacy. Narrative medicine promises to give this.

The impersonality of contemporary health care may have a number of sources, but it is typically blamed on the "biomedical model" which treats diseases as kinds of biological dysfunction or on "evidence-based medicine" which reports statistical results. This is then contrasted with "humanistic" medicine, whose methods come from the humanities, which focus on concrete events and experiences rather than on generalizations and abstractions about them. The implication is that the humanities are more humanistic than the sciences and that in order to humanize medical practice we need to have more humanities expertise. This puts the medical humanities in a favorable light and provides widely used justifications for their inclusion in the medical curriculum.

I think that the disciplinary differences are not really the issue here. The humanities are not always humanistic[23] (e.g. much of narrative theory is ethically disengaged), and technoscience is not always impersonal (e.g. artificial limb construction and adjustment is carefully adapted to the specific bodies and needs of disabled individuals). Impersonality may come more from provider fatigue and the bureaucratic structure of health care than it does from the biomedical model or from the influence of evidence-based medicine.

---

[23] One might think that it is analytic (a necessary truth) that the humanities are humanistic. But this is so only if "humanistic" is taken to mean "of the humanities." My sense of the usage of "humanistic," especially in medical contexts, is that it is used to mark ethicality. Marya Schechtman suggested to me that "humanistic" includes a wider range of issues, including aesthetics. If "humanistic" has this broader meaning, and the wider range of issues is described in terms of humanities disciplines, then it is analytic that the humanities are humanistic. Then I would drop my claim that the humanities are not humanistic, but continue to argue that technology can be humanistic.

Respect for the individuality of the patient is not the only way in which narrative medicine is personal. The role of the physician is also described personally and individually, since he or she is personally engaging with the patient. The humanity of the physician as well as of the patient is part of the ethos of narrative medicine. The bureaucratic and increasingly regulatory structure of health care is a problem for many physicians as well as for patients. Narrative medicine is a practice that respects physician individuality as well as patient individuality. And it is a short step from respecting individuality to respecting autonomy. Narrative medicine is also a practice that respects physician autonomy as well as patient autonomy. Moreover, because it is common to think of humanity in terms of identity, and identity in terms of uniqueness, narrative medicine is seen as a practice that respects both physician and patient humanity.[24]

Physicians regard themselves as members of a profession rather than a trade. The sociologist Eliot Freidson has argued that professions are characterized by their insistence on the discretion, control, and autonomy of the individual practitioner (Freidson 2001).[25] The empathic engagement of physician and patient is also private (as well as confidential), not easily accessible for judgment by others. This can function to maintain the professional authority of the physician viz-à-viz regulators such as governments and insurers. Thus the ethos of narrative medicine also supports individual physician authority. (This is not the authority of the physician over the patient, but the authority of the physician's actions as a professional.)

The ethos of narrative medicine is quite different from that of "the biopsychosocial model" (Engel 1977) that was popular in the 1970s and 1980s, but which has now almost disappeared. The biopsychosocial model was an early attempt to expand on the biomedical model by adding psychological and social factors to accounts of health and disease. The result was a complex (but otherwise unspecified) model in which physical, psychological, and social factors interact with one another causally. But the biopsychosocial model lacks the intimacy, empathy, democracy, and dyadic connection that is so central in narrative medicine. I think that

---

[24] I do not accept all these equations. In particular, I do not accept that our identity or our humanity lies in our uniqueness alone. These equations are common in Western thought, however, especially in the US.

[25] I thank Alan Richardson for bringing the work of Eliot Freidson to my attention.

narrative medicine has followed (indeed, often replaced) the biopsycho-social model because of its ethos, which is widely appealing. I will say more about the differences between narrative medicine and the biopsy-chosocial model in Section 8.4.

One way of characterizing the ethos of narrative medicine is to say that it is the counter to "cookbook medicine." "Cookbook medicine" is a term of criticism used in objections to consensus conference, clini-cal guidelines, and evidence-based medicine. The implied criticism is that the medical practice is "one size fits all" rather than nuanced to individual cases and also (to my ear) that those who practice "cookbook medicine" are not really thinking about it or taking its human implica-tions seriously (just as a line cook may mindlessly follow instructions). Perhaps the term "cookbook medicine" is also suggesting that those who practice it are mere tradespeople (like cooks) rather than profes-sionals (like doctors).[26] Again, I think the real target of this criticism is bureaucratic medicine, not consensus conferences or evidence-based medicine.

## 8.4  Appraisal of Narrative Medicine

In this section I make some general comments about, as well as criticisms of, narrative medicine.

### 8.4.1  "Narrative" Has Broad Scope

The term "narrative," as used in "narrative medicine," has broad scope. It includes the content of stories, the literary analysis of stories, the abil-ity to listen to stories, the empathic comprehension of stories, nonver-bal communications, the experiences of patients and physicians, and the creative ability to tell new narratives. Narrative competence for physicians has four components, as discussed in Section 8.2: listening, empathizing, narrative detective work, and making meaning. The vast majority of examples of narrative medicine in the literature are examples in which patients have their psychological needs addressed in the con-text of a caring relationship with the physician. There are a few examples of use of the technical tools of narrative theory for narrative detective

---

[26] The term "cookbook medicine" unfairly insults cookbooks. It also insults cooks, suggesting that they are less than professionals.

work, and a few examples that deal with physical (non-psychological) needs alone.[27]

It is helpful to ask the question: "What is *not* included in narrative medicine?" Most of those who write in the field of narrative medicine accept the traditional dichotomous framework of an art and a science of medicine, making narrative medicine the art and "the biomedical model" or "evidence-based medicine" the science. Moreover, especially in the writing of Rita Charon, narrative medicine is thought of as concerned with particulars and the science of medicine as concerned with generalizations (concerned with the laws of nature). Narrative medicine is then commonly contrasted with evidence-based medicine (which gives statistical generalities), which makes the challenge for the clinician to "integrate narrative and evidence-based medicine." Books like James Meza and Daniel Passerman's *Integrating narrative medicine and evidence-based medicine* (2011) are the result, and they suggest that evidence-based medicine provides the general background knowledge and narrative medicine does the work of tailoring this knowledge to particular cases.

This tidy view cannot be quite correct, since some of the work of tailoring general knowledge to particular cases cannot be described as "narrative" work. In particular, as was discussed in Chapter 6, evidence-based medicine is applied in particular cases by using causal and mechanistic reasoning about some of the specifics of the cases. An example would be a decision not to give beta blockers for migraine prophylaxis to a patient with naturally low blood pressure. This kind of reasoning used in tailoring to specific cases is not thought of as "narrative reasoning"; it is the same kind of local application of scientific reasoning as is used in, say, weather forecasting or mechanical engineering. There is more to non-narrative reasoning than evidence-based medicine, and more to non-narrative reasoning than pathophysiological generalities. Application of general pathophysiological knowledge to specific cases does not typically need narrative expertise; what it needs is local knowledge of specific cases.

---

[27]  Why aren't there more examples of narrative detective work and non-psychological therapies? I wonder whether the focus on the other kinds of narrative competence, as well as on psychological health, reflects the actual rarity of cases of narrative detective work and physical therapies?

Rita Charon, Kathryn Montgomery, and others characterize non-narrative reasoning as "logico-scientific."[28] For them, scientific and mathematical argumentation is not narrative reasoning, and narrative reasoning is a different kind of argumentation. I have already discussed the ways in which this bifurcation fails because scientific reasoning often makes use of causal narratives. It is more accurate to say that some kinds of scientific reasoning, such as statistical modeling, are not narrative reasoning.

### 8.4.2 Narratives Are not Entirely Singular

In much of the writing on narrative medicine, narratives are cherished for their particularity and for the details about each case. Patients can have the same disease but very different narratives of illness. Narrative medicine promises to treat each patient "as an individual," with attention to their unique identity. This is a heady promise, especially in the culture of the United States and other liberal democracies, which value unique-ness and individuality. But narratives are not completely singular.

At a minimum, published narratives are assumed to keep their iden-tity despite changes to the patient description sufficient to protect the patient's anonymity. All those writing in the area of narrative medicine make these changes. This means that each story is a member of a class of stories that is essentially the same, with differences in identifying details. Indeed, Charon tells the story described in Section 8.2.1—about a 36-year-old Dominican man with back pain (Charon 2004)—almost verbatim in another publication as about a 46-year-old Dominican man with shortness of breath and chest pain (Charon 2006).

Arthur Frank (1995) classifies narratives more generally, distinguish-ing restitution, chaos, and quest (redemption) narratives. During serious illness, patients often express "chaos" narratives which are in due course resolved either by "restitution" narratives which restore the original self or by "quest" narratives in which the self is fundamentally and positively transformed. Common tropes in restitution narratives are gratitude to one's physician, surgeon, or medication for restoring one's health. Common tropes in quest narratives are a new sense that every day counts, a discovery of a deep commitment to faith or family, and the

[28] The term "logico-scientific" originates from Jerome Bruner's *Actual minds, possible worlds* (Harvard, 1986).

emergence of political activism in order to get attention and resources for the disease. Frank claims that individual stories typically mix the three different narrative types in some unique way, thus preserving the overall uniqueness of stories (1995). But this is simply a claim that is made to support the uniqueness of stories, and no evidence is given for it.

In fact, many illness narratives have a similar structure: the hope for simple restitution, which is then dashed, subsequent descent into chaos and despair, and eventual redemption with a quest narrative in which the person's life is enriched. Military metaphors of "fighting" disease are common. In Section 8.2.4 I introduced a standard late twentieth-century breast cancer narrative that I will elaborate here. It begins with finding a breast lump either in mammography or physical examination, continues with the wish that the lump will turn out to be "nothing," moves through the cancer diagnosis and initial despair, finds purpose (usually in others, such as children) to sustain the determination to get through grueling treatment, and ends with the discovery of some personal positive transformation. Moments of humor are often employed to lighten the tone. Betty Rollin's book *First, you cry* (1976) was one of the earliest such narratives, and the cover blurb describes it as "the inspiring, true story about how one woman transformed the most terrifying ordeal of her life into a new beginning." It was brave at that time to write a personal narrative of experience with breast cancer, but after more than 30 years of similar narratives and a prominent redemptive "pink ribbon" movement the narrative has become familiar, even formulaic at times. The experience of breast cancer treatment has become a female rite of passage, one that is now more feminine than feminist in its focus on pink accessories.

Barbara Ehrenreich (2001) and Judy Segal (2007) have both commented critically on this standard breast cancer narrative. Ehrenreich (a writer and activist) castigates the "mandatory cheerfulness" of the breast cancer world for implicitly denigrating the dead and dying, likens breast cancer culture to a religion or cult, and says the following about breast cancer narratives:

The personal narratives serve as testimonials and follow the same general arc as the confessional autobiographies required of seventeenth-century Puritans: first there is a crisis, often involving a sudden apprehension of mortality (the diagnosis, or, in the old Puritan case, a stern word from on high); then comes a prolonged ordeal (the treatment or, in the religious case, internal struggle with the Devil); and finally, the blessed certainty of salvation, or its breast cancer equivalent, survivorhood. (Ehrenreich 2001: 50)

Segal (2007), a rhetorician of health and medicine, goes even further and argues that the ubiquity of similar breast cancer stories and the negative reactions to stories that depart from the norm are evidence of "narrative regulation" that suppresses other kinds of narrative. She claims that the standard story of personal struggle and personal victory is appealing in part because it gives cancer patients agency. It also gives currently healthy but anxious people agency with frequent reminders of their responsibility to get regular mammograms. The standard story may be an improvement on older narratives of physician authority and patient passivity. However, as a recognized narrative type, it has come to play a normative role in discouraging other narratives, such as those that focus on the environmental causes of breast cancer and what the community can do to address them. It also discourages narratives that describe personal struggle and do not end in personal victory. And it discourages narratives that talk about prevention rather than about detection and cure. Segal draws the grim conclusion that pharmaceutical companies and governmental agencies have embraced and repeated the personal breast cancer narrative because it serves their interests by distracting people from paying attention to the environmental causes of breast cancer.

Even Frank's framework of restitution, chaos, and quest incorporates some narrative regulation (although not as much as the standard breast cancer narrative). It permits negative elements only in the form of "chaos"—moreover, chaos that begs to be resolved in a restitution or quest narrative. Non-chaotic negative processes, negative conclusions, and even chaotic conclusions are not among Kleinman's narrative templates. It is possible that chaotic conclusions may not be in any narrative template, if narratives must serve to make order from chaos. But do narratives always make order from chaos? Shlomith Rimmon-Kenan (2006) argues persuasively that narrative theory should learn from the chaos of illness narratives that the imposition of structure is sometimes impossible.

I think that the belief that narratives are singular gets in the way of discovering and acknowledging that narratives often have structural similarities with one another, and prevents us from seeing that the structural similarities can serve various purposes, including political purposes. Belief that narratives reflect or express personal identity or personal uniqueness also discourages political and evaluative analysis; the ethos of respect for patients' personhoods gets in the way of acknowledging the role of narrative regulation.

There is, in addition, a general tension between identity and individuality in these discussions. Identity is often expressed in terms of membership of particular social groups (such as feminists or Catholics or bisexuals) but individuality is often portrayed as singularity and uniqueness. How much of each can we have? The more we have of one, the less we have of the other.

### 8.4.3 Narratives Can Contain Falsehoods, and Partial Truths

There is a surprising credulity in the narrative medicine literature about narratives, especially patient narratives. The only remark Rita Charon makes about this is that "we enter others' narrative worlds and accept them—at least provisionally—as true" (2006: 10). In fact, Charon and most who work in the area of narrative medicine give no examples in which patients misremember, fabricate, exaggerate, lie, distort, selectively tell, or otherwise intentionally or unintentionally, explicitly or implicitly, report falsehoods, perhaps in an attempt to present themselves in a favorable light to their physician.

John Hardwig's paper, "Autobiography, biography and narrative ethics" is a notable exception which remarks, "The epistemic and moral weaknesses of autobiography are obvious and commonly recognized.... One interesting question, then, is why narrative bioethics has not already recognized them" (1997: 51).[29] Hardwig goes on to argue that a combination of the "ghosts of Cartesianism" which regard first person reports as trustworthy and the "much too simple patient-centered ethics," which privileges the patient's story, are responsible for the failure to recognize the weaknesses of autobiography. I am not convinced by the first part of Hardwig's analysis. Narrative medicine theorists are not generally rooted in Cartesian thought, although many philosophers are. The second part seems closer to the mark and is worth developing. Distrust or skepticism on the part of the physician can damage the intimacy of the dyadic patient–physician encounter that is at the heart of narrative medicine. Moreover, when the patient's narrative is identified with the patient's self, any skepticism seems additionally disrespectful and thus contrary to the ethos of narrative medicine.

It is well known in the medical profession that people sometimes lie (or misremember, fabricate, distort, selectively tell, self-deceive, or

---

[29] Hardwig is writing about narrative ethics specifically, but his point applies to narrative medicine more generally.

otherwise intentionally or unintentionally misrepresent) about matters such as substance abuse, sexual activity, exercise habits, and so forth. Humans are also in general susceptible to the "narrative fallacy" in which the attempt to weave experience into a coherent story results in the omission of facts, or even in their (intentional or unintentional) distortion or fabrication. Patients also seek narratives in order to answer their questions about illness and accept those narratives that most satisfy their psychological needs, rather than those that are more likely to be true. For example, Julia Ericksen, in *Taking charge of breast cancer* (2008), finds that women who get breast cancer invariably have a story about why they got breast cancer, yet these stories say more about the personalities and needs of the patients than they do about the likely causes of or actual risk factors for their disease. For example, "traditional responders" to a breast cancer diagnosis tend to blame themselves in some way (for not having regular mammograms, or for having an abortion or a miscarriage); "alternative experts" tend to blame something in the environment (such as proximity to nuclear plants or secondhand smoke).

Even when narratives are true, they may only be part of the truth. Coherent narratives can have an illusion of completeness in which all is apparently causally accounted for, missing the possibility that some things have multiple causes. This is of course why different (but not inconsistent) narratives of the same event can sometimes both be true.

### 8.4.4  Singularity Is not Intimacy

We associate intimacy with the sharing of details about our lives. People share more details with intimate others. Most of us deeply appreciate intimate relationships in which we can share the prosaic details of our days. But the sharing of details does not in itself produce intimacy, as anyone who has been stuck with a bore at a party can testify. It is more that the presence of intimacy makes the sharing of details more likely and more interesting. The sharing of details is neither necessary nor sufficient for intimacy.

### 8.4.5  The Health Care Team

The doctor–patient dyad is part of the ethos of narrative medicine. It is certainly attractive to many people—both patients and physicians—to have this couple at the core of health care. I already remarked that the individuality of both the patient and the physician are emphasized in the doctor–patient dyad. We regard two person relationships as potentially the most intimate and trusting of all relationships, as exemplified in mother–child bonding and pair bonding (romantic, sexual, and/or reproductive). But for some kinds of health care, the therapeutic dyad may not be the best structure for taking care of patients. For some diseases—for example, cystic fibrosis care and cancer treatment—the patient does best when several health care providers with a range of expertise and roles work together as a team to take care of them. For the cystic fibrosis patient, that team may include pediatric pulmonologist, respiratory therapist, nutritionist, and social worker. For the cancer patient it may include surgeon, oncologist, interventional radiologist, nurses, physical therapist, psychotherapist, chaplain, and patient support group. Regular team meetings serve to keep everyone on the same page.

It is best to keep an open mind about the best social structure for patient care. The doctor–patient dyad is the most traditional therapeutic social structure but our attachment to it may be more sentimental than pragmatic.

Several of Rita Charon's cases describe her sensitive treatment of patients' considerable psychosocial needs. She treats patients in circumstances when other primary care practitioners would refer them to psychotherapists, psychiatrists, or other mental health workers. Clearly, Charon is especially skilled in this area. Many primary care practitioners are not. They are not abandoning their patients if they refer them to specialists in their areas of need. In addition, the fourth narrative competence, making meaning, is one that is often found in non-medical contexts such as philosophy, psychology, literature, and religion.

Cases of narrative detective work notwithstanding, narrative medicine is mostly about patients' psychosocial needs. It is not surprising that those who advocate narrative medicine tend to be primary care physicians, psychiatrists, and medical ethicists. Narrative medicine may be less important in other specialties. In addition, as Charon has often said, narrative medicine is important to other members of the health care team, especially nurses, social workers, and chaplains.

### 8.4.6  Relationship of Narrative Medicine to Narrative Therapy

Narrative therapy is a kind of psychotherapy that was developed by family therapists Michael White and David Epston (1990). It aims to develop patient narratives that empower the patient(s) and encourage positive change. It is more normative than the field of narrative medicine in that it favors empowering narratives over disempowering narratives. Because of the awareness that we have a choice of narrative(s), it is more aware of the constructed nature of narratives. So far as I know, the fields of narrative therapy and narrative medicine developed independently, although they may both have been generally influenced by the "narrative turn" in academia.

### 8.4.7    Relationship of Narrative Medicine to the Biopsychosocial Model

The biopsychosocial model (Engel 1977) was described in Section 8.3. It includes attention to psychological and social well-being. More recently Daniel Sulmasy (2002) has added spiritual well-being to the model. I have already mentioned the shortcomings of the approach: the complex causal model of interaction of biological, social, and psychological factors is unspecified, and the biopsychosocial model lacks the ethos of narrative medicine (which is attractive to many). But there is an important advantage that the biopsychosocial model has over the narrative medicine model, in my assessment. The biopsychosocial model can highlight political structures, such as social class, as causal players in the health of patients. Narrative medicine, with its insistence on individual narratives, often misses more general structures, including political dimensions. For example, if social class is not part of the patient's or the physician's narrative, it is left out altogether.

Much of narrative medicine is politically disengaged. A notable exception is the work of Hilde Lindemann (Nelson 2001). Her work is both narratively creative and politically aware. She actively "makes meaning" rather than passively finds it in the flow of patient and physician–patient dialogue. In its narrative creativity, her work is similar to that of narrative therapists. But while narrative therapists are interested in psychologically empowering narratives, Lindemann is interested in politically empowering narratives. Lindemann uses her expertise as a feminist scholar to do this political work. It is not part of the regular ethos of narrative medicine.

An example (Nelson 2001: 3) is the case of nurse Pilar Sanchez, who took care of a 16-year-old boy who was dying from leukemia. The boy's parents requested that he not be informed of his prognosis. Pilar Sanchez spoke to the oncologist, and argued that the boy should be informed. The oncologist dismissed her request, casting it as the result of "overinvolvement" with her patient. Through discussions with other nurses, Sanchez realized that the oncologist was seeing her as an excitable Hispanic woman and that this "master narrative" was getting in the way of appreciating that Sanchez had developed expertise on the boy's wishes and was in fact supporting the patient's developing autonomy.

## 8.5  Conclusions

The widely felt need to advocate for more humanistic care in medicine is behind much of the literature on narrative medicine. Although I agree that there is such a felt need, I think that narrative medicine is just one way to respond to it, and that its flaws have been overlooked in a general enthusiasm about having a more humanistic medicine. Its flaws include the fact that narratives are easily fabricated and sometimes leave out important data, and the fact that the popularity of particular narrative forms can suppress the formation of other narratives. Moreover, the insistence that narratives are singular tends to discourage important political analyses of the situation of individuals (an exception here is the work of Hilde Lindemann). Finally, blaming "the biomedical model" or evidence-based medicine for the impersonality of medicine is criticizing science for the institutional failings of health care bureaucracy.

This chapter distinguished the concepts of *biological* variability, *psychological* variability, and *personhood*, which are often conflated in the medical humanities literature. In my view, personhood and respect for persons should not rest on the presence of individual differences (physical, psychological, or narrative) that are thought to make persons "unique." I think that we are people at least as much because of our sameness to one another as because of our differences.

Narrative medicine places at its core an intimate dyadic relationship between physician and patient. There is great nostalgia for this kind of healing relationship, which predates the birth of Western scientific medicine and is found in many, perhaps all, cultures. But I do not think that such a healing relationship is our only option, or even our best option, at this time. In some areas of medicine, management by a health care team rather than by an individual clinician is the most effective. Some members of the team—such as psychologists, social workers, and chaplains—have special expertise in the tasks of empathizing and "making meaning," and the patient can even go out of the medical context altogether to address their psychosocial needs.

Narrative medicine was developed by primary care physicians and psychodynamic psychiatrists, specialties which foreground the physician–patient dyad. It is understandable that these providers have a wish to meet all their patients' needs, and that their patients are gratified to have all their needs met by a single caregiver who knows them well and whom they trust. But this is just one therapeutic ideal and a somewhat romantic model of patient care, even for primary care practice and psychiatry. Many primary care physicians are not as sensitive as Charon, and many psychiatrists are not as empathetic as Halpern. With team care models, we may be developing other ideals for comprehensive patient care. With modesty about the goals of medicine, we may also acknowledge that some patient needs can be addressed out of the medical context.

# 9

# A Developing, Untidy, Methodological Pluralism

## 9.1 Introduction

In previous chapters I explored consensus conferences, evidence-based medicine, translational medicine, and narrative medicine in turn. All the methods are in active use; the situation is one of methodological pluralism.[1] This chapter looks at how the methods relate to each other. A first approximation view of the methods is that they fall on a spectrum between clinical research methods on the one end, and the clinical encounter on the other end. Translational medicine is the first step of clinical research (the lab through Phase I and II trials), then when a promising intervention is found the methods used to ascertain that promise are those of evidence-based medicine (Phase III trials, meta-analyses, and systematic review), after which the results are disseminated by consensus conferences and implemented in the clinical encounter using narrative medicine. On this view, each method has its place and its role, and no method need come into conflict with another method because the methods answer different questions. This first approximation view is misleading and incomplete. Like the traditional art–science dichotomy, its simplicity is attractive but worth resisting because there is much to learn from the ways in which the methods do not fall on a tidy linear spectrum.

---

[1] Pluralism is a popular term these days in philosophy of science. Most often, it is used to describe pluralism about theories (ontological pluralism), for example in the work of Cartwright, Dupré, Giere, Longino, and Mitchell. Methodological pluralism (pluralism about methods and methodologies) has not received as much attention, but it is present in the work of Bechtel and Richardson, Feyerabend, Kuhn, Longino, and Wimsatt.

Sometimes the methods are used at different stages of inquiry, and do not conflict. But sometimes the methods are used at the same time, to answer the same questions, and then they may agree or disagree about how to answer the question. Here are illustrative examples of these three different kinds of interaction.

Questions about genomic and proteomic therapy for cystic fibrosis are currently addressed within translational medicine, and not by the other methods. This is because genomic and proteomic therapies for cystic fibrosis have not yet worked well enough to have reached the stage of evaluation in large randomized controlled trials (Phase III trials), let alone consensus guidelines for practice. Narrative medicine and the medical humanities have not been used in working out the clinical details of the therapies, although stories such as Emma Whitehead's amazing recovery (in Section 7.3) do contain important reflections on topics such as hope. This is an example in which the methods do not yet interact at all, although they are expected to interact in the future as, for example, translational medicine results are used to design Phase III trials and then the results of Phase III trials are systematically evaluated before consensus clinical guidelines are devised.

Questions about the treatment of newly diagnosed Type 1 diabetes are currently addressed in a unified and thorough way by all the methods. This is because the pathophysiological (causal, mechanistic) understanding of Type 1 diabetes as insulin deficiency (from basic research) has led to the development of effective insulin replacement therapies (from translational medicine to evidence-based medicine), and there are standard guidelines for treatment in team-based settings that can be adjusted in response to the patient's psychosocial needs. (For example, different kinds of insulin delivery systems can be used.) These standard guidelines are produced regularly by expert consensus processes based on the latest evidence.[2] This is an example in which the different methods each have a clear role and do not come into conflict with one another.

Screening mammography for women aged 40–49 is an area in which the methods produce some conflicting results. While most evidence-based analyses do not find that benefits outweigh costs, and

---

[2] See, for example, the American Diabetes Association's (ADA's) "Standards of Medical Care in Diabetes" revised annually by the ADA's Professional Practice Committee, based on the evidence: <http://care.diabetesjournals.org/content/36/Supplement_1/S11.full>, accessed July 26, 2013.

therefore advise against screening mammography for women aged 40–49, several consensus statements (such as those of the American Cancer Society and the American College of Radiology), and many patient and physician narratives, continue to insist that the benefits outweigh the costs and urge that women aged 40–49 continue with annual screening mammograms. In this chapter I will look at this case in detail. There is much to learn from it about how the methods can produce conflicting results, and how controversy is managed in medicine.

These three examples—genomic and proteomic therapies in cystic fibrosis, management of Type 1 diabetes, and the use of screening mammography for women aged 40–49—illustrate the three different ways in which methods can relate to one another. They can work in independent domains. They can work in overlapping domains, delivering results that either agree or disagree with each other.

If the methods always worked in different domains or for different questions, we would have what I call a "tidy pluralism"; each method could have its own domain of application coinciding with its domain of authority. This is the situation, for example, in Stephen Jay Gould's popular account of the "non-overlapping magisteria" of science and religion.[3] Instead, we have a situation in which sometimes the methods apply to answer the same questions, giving sometimes the same and sometimes different answers, and there is no "meta-method" to which to appeal to resolve the disagreements. I call this situation "untidy pluralism." Moreover, we expect new and hopefully better methods to develop in the future, so the current untidy pluralism is expected to change in its details over time. Hence the long title of this chapter: a developing, untidy, methodological pluralism is the most accurate description of the position (it needs an acronym, but the most obvious one—DUMP—does not have the right kind of gravitas).

Methodological pluralism is one source of disagreements in medicine, but it is not the only source. There can also be disagreements among different practitioners allegedly using the same method or methods. For example, some clinical trials fail to replicate. When a trial is repeated and the results are different, the overall evidence is unclear, and disagreement

---

[3] See <http://www.stephenjaygould.org/library/gould_noma.html>, accessed July 26, 2014

is usually the result. (Sometimes suspension of judgment is the result, especially in bystanders.)

Meta-analysis is a tool for combining the results of several trials to see if there is a discernible consistent interpretation of the results. So meta-analysis is potentially a tool for reducing disagreement resulting from different trial results. However, sometimes those using evidence-based methods to do systematic analyses or meta-analyses of the same body of evidence produce results that disagree with one another—see, for example, the case of different meta-analyses of the results of mammography in Goodman (2002). This happens because meta-analytic methods are not algorithms, and there is a need for what is often called "judgment" in how they are applied. Experts do not always make the same judgments.[4]

Disagreement is not, *ipso facto*, an undesirable state of affairs. Some time ago I wrote a book about how dissent is underrated by philosophers and historians of science (Solomon 2001). Dissent promotes healthy criticism, produces useful division of intellectual labor, and is a predictable outcome of useful diversity of skills and training. The meaning of dissent is not that "we need to find out who is right and who is wrong" but rather that "more than one theory may have some truth in it." There can be scientific progress during times of dissent, such as when competing theories are developed and tested.

That said, disagreement in science can have a downside. When science is used in applications such as medicine, climate policy, or toxicology, disagreement among experts tends to decrease the authority of those experts, and this loss of authority can be exploited by special interests such as medical insurance companies, climate change deniers, and corporations. Indeed, this is the flip side of the rhetorical importance of consensus conferences (discussed in Chapters 2, 3, and 4); when there is a lack of consensus among experts, the experts lose some authority. So, while dissent in science is generally beneficial for the scientific enterprise, dissent in applied sciences, including medicine, can lead to scientists' pronouncements being taken less seriously. Logically speaking, scientific disagreement should not make the claims of cigarette companies,

[4] The term "judgment" suggests that differences in assessment will be rational rather than arbitrary. In this way, "judgment" is used normatively rather than descriptively. I'm reporting this usage, not agreeing with it.

climate change deniers, or pharmaceutical companies any stronger than they actually are. The proper response to scientific disagreement is to take all the *scientific* positions seriously, and, if you are a scientist, support further development for each of them. But practically and rhetorically speaking, the response to scientific disagreement has been to give more credence to other interests. Thus, if scientists want to be heard in a public context it is in their interest to advertise their agreements rather than their disagreements. Hence the consensus statements of the IPCC are important not for science but for the epistemic authority of science in public discourse.

It is, therefore, not surprising to find strategies and processes that "manage" disagreements in areas of science that are in the public eye, either by avoiding the disagreement in the first place or by containing the scope of disagreement. Overt disagreement in medicine is less common than it might be because of these strategies and processes. The screening mammography case to be discussed in this chapter is atypical because the disagreements are not completely managed away, making it a useful case for identifying the social epistemic phenomena at work.

I present the case of screening mammography for women aged 40–49 with as much "methodological equipoise"[5] as I can muster. I am not trying to come to any conclusions about which method or which position is correct. My goals are to analyze how the disagreements arose, how they are managed, and what this tells us about methodological pluralism in medicine.

## 9.2   Case Study: Controversy over Screening Mammography for Women Aged 40–49

Recommendations for screening mammography have been controversial ever since the technology was developed in the late 1960s.[6] Despite the fact that mammography has been more extensively evaluated by randomized controlled trials than any other screening method (Wells 1998), it continues to be controversial, especially for routine use in women

---

[5]   This term was introduced and discussed in Chapter 1.
[6]   This case study was first presented in Solomon (2012) and there are occasional overlaps of text.

aged 40–49. The length and persistence of controversy—over 40 years, with the same issues frequently rehashed—is extraordinary. The Canadian Task Force on Preventive Health Care selected screening mammography as its first topic for discussion after its re-establishment in 2010, indicating the continuing importance of the topic. As of July 2014 in the US,[7] the American Cancer Society, the National Cancer Institute, the Society for Breast Imaging, and the American College of Radiologists all recommend annual screening starting at age 40, while the USPSTF and the American College of Physicians recommend against routine mammography for women aged 40–49.

Judgments on the topic differ pointedly. Michael Baum, a breast surgeon in the United Kingdom, asserts that screening mammography is "one of the greatest deceptions perpetrated on the women of the Western World" (quoted in Ehrenreich 2001: 52), while the American College of Radiology has a statement on its website about the "ill advised and dangerous USPSTF mammography recommendations" which claims that "The [USPSTF] recommendations make unconscionable decisions about the value of human life" and predicts that the recommendations will lead to more women dying of breast cancer.[8] In response to such comments, Jane Wells, a British public health physician, astutely states: "The debate over the necessity of screening for breast cancer among women in their 40s has assumed an importance out of proportion to its potential impact on public health" (Wells 1998: 1228).

Quanstrum and Hayward (2010) have analyzed the ongoing controversy in terms of the self-interest of radiologists. It is plausible that the recommendations of the Society for Breast Imaging and the American Society of Radiologists were at least partly influenced by the desire to provide services in screening mammography. But the epistemic situation is more complex, both for the radiologists and for other participants in the controversy.

In Section 9.2.1, I argue that methodological pluralism is responsible for a good deal of the controversy. At least four different methods

---

[7] There is similar controversy in other developed countries, but I am most knowledgeable about the US.

[8] <http://www.acr.org/About-Us/Media-Center/Position-Statements/Position-Statements-Folder/Detailed-ACR-Statement-on-Ill-Advised-and-Dangerous-USPSTF-Mammography-Recommendations>, accessed July 26, 2014.

are used in making judgments about the value of screening mammography: pathophysiological (causal, mechanistic) reasoning, evidence-based medicine, consensus of experts, and "clinical judgment" (which includes narrative medicine). I go through these in turn to see what they deliver.

### 9.2.1 Pathophysiological (Causal, Mechanistic) Reasoning

When mammography was invented, cancer was understood as a progressive disease in which malignant cells multiply at first locally and then distantly (as metastases). The prognosis was thought to be better the earlier that the malignant cells are discovered and then eradicated by surgery, radiation, or chemotherapy. The mantra "early detection saves lives" has been widely embraced and has evidence in its favor, especially for some types of cancer such as melanoma, colon cancer, cervical cancer, and testicular cancer. According to this reasoning, screening mammography is expected to save lives, since it detects cancers too small to be felt in a clinical breast examination, and thus at an earlier stage of development. This expectation was so strong that when in the early days of the Canadian National Breast Screening Study the data appeared to show that 40–49 year old women had *more* deaths from breast cancer than controls who were not screened, the results were dismissed as a statistical fluke (Baum 2004).

Since the 1960s and 1970s, we have learned that cancer is a more complex class of diseases, and that early detection does not always save lives because even in early stages cancer can be a systemic disease. Prognosis is affected by the tissue of origin, the particular genetic mutation, and the host immune response to the cancer. Michael Baum et al. (2005) have extended the angiogenesis work of Judah Folkman, and suggested that mammography *causes* breast cancers through the unforeseen effects of biopsy of suspicious lesions. Baum suggests that the injury caused by biopsy produces increases blood flow and stimulates angiogenesis in the area, and that this nourishes the cancer and enables it to spread. In the current state of knowledge about breast cancer and its mechanisms, this is a reasonable hypothesis. We do not have any evidence for the hypothesis at this point. But recall that we do not have clear evidence that screening mammography saves lives in women aged 40–49 either.

## 9.2.2  Evidence-Based Medicine Results

Nine randomized controlled trials of screening mammograms have been performed in the United States, Canada, and Europe, beginning in the early 1960s. Most of the trials, especially in the early years, have methodological flaws, ranging from post-randomization exclusions to inadequate randomization to unblinded cause of death attributions (Goodman 2002). The most recent trial, called the Age Trial—conducted in the United Kingdom, begun in 1991 and reported in 2006 (Moss et al. 2006)—was designed to look specifically at the question of the efficacy of mammography in women under 50. In all the studies, any benefit of mammographic screening is small, even for women aged over 50.

The effect found in the most recent Cochrane review (Gotzsche & Nielsen 2011) is a reduction in breast cancer mortality of about 15 percent for women *of all ages* who undergo regular screening mammography. This may sound substantial, but absolute risk reduction is very small: 0.05 percent. (This means that a woman who has a 5 percent chance of breast cancer mortality now has a risk of 4.95 percent.) Absolute risk reduction is even smaller in women aged 40–49, who have a lower chance of dying from breast cancer than older women. The Cochrane review estimates that approximately 2,000 women over the age of 50 need to be screened annually for ten years to prevent one "premature death" from breast cancer; this averages out to a life extension of two days for every woman who undergoes ten years of mammography.[9] (Using the latest Cochrane review as a reference point skims over some of the controversy; other meta-analyses have come to different conclusions. See Goodman (2002) for a discussion of this.)

The Age Trial, specifically designed to look at the effect of mammography in women in their forties, did not find a significant difference in breast cancer mortality between the screened and the unscreened groups (Moss et al. 2006). The actual difference was in the expected direction

---

[9] Since getting a mammography takes time, and that time is usually not enjoyable, the question about whether or not to get one could be framed as: "Is it worth it to you to spend time getting a mammogram when the average benefit of that mammogram is an additional 4–5 hours of life?" Or it could be framed differently, as: "Is it worth it to you to spend time getting a mammogram in the extremely unlikely event that it will prevent your premature death?" Decisions are sensitive to framing effects.

214 A DEVELOPING, UNTIDY, METHODOLOGICAL PLURALISM

(screened women had a slightly lower breast cancer mortality) but it was not statistically significant.

Screening mammography has possible harms as well as possible benefits. These harms include the psychological harms of the many false positives (around 200 for every life saved), the physical harms of the additional radiation and surgery needed to follow up the false positives, and the psychological and physical harms resulting from positive diagnoses and treatments of cancers that would otherwise not have been noticed. Peter Gotzsche and Ole Olsen (2000) published an analysis showing that breast cancer screening—for women of all ages—produces no *overall* reduction in mortality.[10] So presumably any gains in breast cancer mortality due to screening are offset by the harms of screening. Morbidity—physical and psychological—should also be considered. But morbidity is difficult to measure. Decisions about overall harms and benefits of screening mammography are sensitive to the magnitudes of harms caused by anxiety and unnecessary treatments.

Any screening technology with harms and benefits will have a clear area of benefit, a clear area of harm, and a gray area (Quanstrum & Hayward 2010). Screening mammography for women aged 40–49 falls in this gray area (and the gray area may extend to screening mammography for women aged 50 and above).

## 9.2.3 Consensus of Experts

There are several consensus statements on screening mammography, some recommending it for women aged 40–49, others recommending against it. As of July 2014, the NGC lists seven guidelines:[11] the Canadian Task Force on Preventive Health care updated guideline from 2011, the US Preventive Services Task Force guideline from 2009, the Kaiser Permanente guideline from 2010, the American Society of Breast Surgeons guideline from 2011, the American College of Obstetricians and Gynecologists updated guideline from 2011, the American College

---

[10] Their conclusions are controversial.

[11] My source for these positions is the National Guidelines Clearinghouse, accessed June 14, 2014. This list is different from the list of recommendations by professional groups at the beginning of this section because the background to making the professional group recommendations is not known to me. They may or may not be based on consensus of experts. The National Guidelines Clearinghouse clearly requires a consensus of experts (as well as an evidence review).

of Radiology guideline from 2012, and the University of Michigan Health System guideline from 2012. The first three guidelines recommend against annual screening for women 40–49, the second three guidelines recommend it, and the University of Michigan Health System guideline straddles the fence but leans toward recommending it.[12] Such consensus statements are typically produced by committees of 10–20 people in face-to-face meetings.

There was even an NIH Consensus Development Conference in 1997 on precisely the question of recommendations for screening mammography in women aged 40–49. Its initial conclusion was that such screening offers no benefits, which surprised the planners of the conference. Two of the panelists changed their minds after the conference following the ensuing outcry in the media from various professional groups. The written record shows a minority opinion dissenting. It is very rare for NIH Consensus Development Conferences to fail to reach unanimous agreement (it has happened only three times in over 150 such meetings). However, it is also well known from the early days of NIH Consensus Development Conferences that they are not good at resolving controversy; they are much better at reflecting general agreement (Wortman et al. 1988).

### 9.2.4  Clinical Judgment and Narrative Medicine

After the 1997 NIH Consensus Development Conference in which consensus was not sustained, John Ferguson (at that time the Director of the NIH Consensus Development Conference Program) wrote an interesting piece about the "great divide" between what he calls "curative medicine" and what he calls "population medicine" (1999). "Curative medicine" is the perspective of the clinician who treats patients individually. "Population medicine" is the perspective of the epidemiologist and evidence-based medicine researcher. Ferguson claims that those who have different perspectives have difficulty understanding each other. In particular, he thinks that clinicians and the public have difficulty with understanding the mathematics of statistics and probability theory.

---

[12] I say that the University of Michigan Health System guideline leans toward recommending it because it interprets the US Preventive Services Task Force guideline as a recommendation that women aged 40–49 talk with their doctor about screening. The US Preventive Services Task Force guideline does say that, but it also says that it recommends against such screening.

He also thinks that clinicians and the public focus on sick people while epidemiologists look at healthy populations. Ferguson observes that the two panelists who declined to endorse the consensus statement are practicing clinicians. (According to the published consensus statement, they are a radiologist and an obstetrician/gynecologist.[13])

Ferguson makes a point that has been made in a number of different ways, and usually from the perspective of the clinician. My own primary care physician believes that epidemiologists are young Washington bureaucrats without clinical experience who are not in a position to make judgments about how to take care of individual patients.[14] Her view (perhaps without the claims about youth and government) is shared by many clinicians. "Clinical judgment" is a common although somewhat vague term that generally refers to particular judgments about actual clinical cases, based on training and extensive clinical experience. It makes use of specific details of cases, as well as their narrative form. Clinical judgment regarding screening mammography takes into account risk factors such as family history and age of menarche, and is also influenced by past clinical experience. Narratives of salient cases, in which screening mammography detected a cancer that was subsequently treated, play an important role for clinicians, most of whom were trained at a time when the "early detection saves lives" mantra was governing and can easily recall a successfully treated and grateful patient.

Patient's perspectives often feature narratives in which screening mammography is given a life-saving role. Many women whose breast cancer is detected by mammography are grateful for the early detection of disease and credit screening mammography with saving their lives, even though such narratives are true for only about 13 percent of them.[15]

Most often, clinical judgment and patient perspectives are sources of arguments for recommending screening mammography for women aged 40–49. I have not come across cases in which narratives are used to argue against screening mammography, even though a clinical experience that could support such a narrative is easy to imagine. Here's a hypothetical scenario: the clinician has several patients in the 40–49 age

[13] See <http://consensus.nih.gov/1997/1997BreastCancerScreening103html.htm>, accessed July 26, 2014.

[14] Conversation with Dr. Margaret Lytton, December 14, 2011.

[15] A recent article (Welch & Frankel 2011) estimates that the chance that mammographic detection of a particular cancer has saved a life is around 13 percent.

group who receive false positive screening tests. They undergo expensive work-ups and take up the clinician's time with requests for reassurance, explanations of risk, and further tests. They show considerable distress during the period of uncertainty and an ongoing fear of screening mammography. The clinician has no experience (or maybe just one experience that can be framed as the exception to prove the rule) of true positive screening tests in women aged 40–49. I expect that there is such a clinician out there—probably several—but we are not hearing from them in public, perhaps because such a narrative could easily be interpreted as showing a lack of patience or care on the part of the physician, and/or make the physician vulnerable to malpractice claims in the extremely unlikely event that a false positive diagnosis turns out to be wrong.

Clinical judgment has a number of well-known biases, such as salience bias (the frequency of salient cases, such as cancers apparently caught by mammography, is often overestimated), the aggregate fallacy (belief that one's patients are atypical, which leads to overruling clinical practice guidelines), commission bias (better to do something than to do nothing), and the tendency to overlook iatrogenic harms. Narrative reasoning also leads to salience biases. So although clinical judgment tends to support routine mammography for women aged 40–49, there are good reasons to be skeptical of conclusions fraught with such potential biases.

While methodological pluralism is responsible for much of the controversy over screening mammography, it is not responsible for all of it. Additional factors, especially in the US, are the atmosphere of litigiousness, which encourages primary care physicians to err in the direction of more interventional testing (and thus more screening mammography), and the political sensitivity of any proposed reductions in spending on breast cancer, which was the focal disease of the women's health movement.

## 9.3 Case Study Continued: Managing Controversy about Screening Mammography for Women Aged 40–49

As stated in the introduction to this chapter, dissent in the medical community tends to undermine medical authority. It is thus in the interest of the medical community to keep dissent to a minimum. The case of screening mammography is unusual: it is a case in which dissent could not be completely avoided or managed away. Vivian Coates (who is the ECRI Institute's Vice President for Information Services and Health Technology Assessment and oversees the National Guidelines Clearinghouse under contract with AHRQ) describes the situation over recommendations for screening mammography as "a troubled area."[16] In the clinical medical context, disagreement means trouble.

To understand the situation better, I ask two questions. First, could there have been *more* dissent over screening mammography? That is, is the current controversy actually less of a controversy than it might have been, because of (active and/or structural) efforts to contain the controversy? And second, what are the differences (if any) between the case of screening mammography and cases in which there is consensus on clinical guidelines? Does the case of screening mammography reflect an extreme of the usual processes, or a different set of processes at work?

I have found three ways in which dissent is reduced or minimized in the case of screening mammography, and one important difference between mammography and many cases in which there is consensus on clinical guidelines. I will describe these in turn.

### 9.3.1 Reframing

One strategy is to reframe areas of dissent positively, as areas for clinical judgment and patient preferences. Clinical judgment includes consideration of patient variability with respect to both benefits and harms and is thought of as part of physician autonomy. Patient preferences include any reasons (good or bad) that a patient uses in making a choice and are thought of as part of patient autonomy. Both physician and patient autonomy are valued, especially in the US.

Guidelines on screening mammography may differ about whether or not they give a general recommendation for or against screening for women aged 40–49, but those that generally recommend against screening also state explicitly that women in this age group may or should make an individualized decision with their physicians. For example, the US Preventive Services Task Force gives screening mammography for women aged 40–49 a "C" rating, meaning that "The USPSTF recommends

---

[16] The comment "a troubled area" was made in an e-mail communication with me, September 21, 2010.

against routinely providing the service. There may be considerations that support providing the service in an individual patient. There is moderate or high certainty that the net benefit is small." However, it uses this to make a more neutral recommendation: "The decision to start regular, biennial screening mammography before the age of 50 years should be an individual one and take patient context into account, including the patient's values regarding specific benefits and harms."[17]

This strategy turns the area of epistemic uncertainty into an area for professional and personal judgment as well as patient-centered medicine. It also, subtly, encourages confining the area where personal judgments and preferences make a difference to women in their forties, because individualized decisions are not encouraged for women over 50. But personal judgments and preferences could reasonably be used for women in their fifties and beyond, to weigh the benefits and risks for each woman. For example, for some women the anxiety associated with false positive results may be severe, and for them the harms of mammography may be greater than the benefits. The magnitude of *benefits and risks* is not the same as the magnitude of *uncertainty in the data*. However, an expansion of patient and physician discretion to women over 50 is not encouraged by any party in the debate.

In my view, there is a convenient compromise here between the desire to be guided by the evidence and the desire to acknowledge both clinical judgment and patient self-determination. Quanstrum and Hayward give an example of endorsement of this strategy when they write of "a gray area of indeterminate net benefit, in which clinicians should defer to an individual patient's preferences" (2010: 1077). The strategy is not limited to the case of mammography; a similar "gray area" has been identified in the use of the PSA (Prostate Specific Antigen) test for routine prostate cancer screening in men aged 55–69, where, again, the US Preventive Services Task Force slightly hedges its generally negative recommendation by allowing "shared decision making that enables an informed choice by patients" in this specific case.[18]

[17] <http://www.guideline.gov/content.aspx?id=15429&search=screening+mammography>, accessed July 28, 2013

[18] <http://www.uspreventiveservicestaskforce.org/prostatecancerscreening/prostatefinalrs.htm#summary>, accessed July 28, 2013.

### 9.3.2  Promoting Trust

It is common for guidelines that do not recommend routine-screening mammography for women aged 40–49 to insist, at the same time, that such mammography continue to be covered by health insurance or government reimbursement programs. For example, the 1997 NIH Consensus Development Conference statement says, "For women in their forties who choose to have mammography performed, the costs of the mammograms should be reimbursed by third-party payers or covered by health maintenance organizations so that financial impediments will not influence a woman's decision."[19] Such a statement is designed to prevent dissent from being amplified by distrust.

### 9.3.3  Producing Guideline Syntheses

The National Guidelines Clearinghouse, managed by the AHRQ, lists up-to-date (within the past five years) clinical guidelines that are produced by recognized organizations and include an examination of relevant scientific evidence.[20] Sometimes this results in listing guidelines whose recommendations conflict with one another. In these cases the National Guidelines Clearinghouse may publish a "guideline synthesis" that describes the areas of dispute and then attempts to come to a more nuanced conclusion that acknowledges the rationales of the different guidelines. I claim that this is partly done in order to restore the guidelines' authority. It may also be done in order to give clinicians more information about the issues at stake.

In late 2010 I looked at the guidelines on screening mammography that were listed on the National Guidelines Clearinghouse at that time (they differ from the listings that I found more recently). They included the American College of Physicians 2007 guideline, the US Preventive Services Task Force guideline from 2009, the American Cancer Society 2007 guideline, and the American College of Obstetricians and Gynecologists 2006 guideline. The first two guidelines were in agreement that routine mammography should not be offered to women aged 40–49 and the second two guidelines were in agreement that it should be

[19]  See <http://consensus.nih.gov/1997/1997BreastCancerScreening103html.htm>, accessed July 28, 2013.

[20]  A full list of past and present criteria for inclusion in the National Guidelines Clearinghouse is stated at <http://www.guideline.gov/about/inclusion-criteria.aspx> accessed October 30, 2014.

offered. However, the NGC Guideline Synthesis, "Screening for Breast Cancer in Women at Average Risk," included only the first two guidelines, thus presenting more consensus than existed at the time. I sent an inquiry to the NGC, asking why other guidelines were listed by NGC (and therefore presumably of high-enough quality) and yet not included in the synthesis. Their response was that the other guidelines are not up-to-date with the evidence. I sent another query, asking why guidelines can be listed (which requires that they be up-to-date with the evidence) and yet not used for the guideline synthesis. This inquiry did not receive a response. Obviously, if selected guidelines are dropped from syntheses, it can look as though there is less dissent than there actually is.

Sometime in 2013, the National Guidelines Clearinghouse updated its material on guidelines for screening mammography. As of July 2014, it lists the seven guidelines I described in Section 9.2.3: three guidelines for, three guidelines against, and one straddling the fence on recommending screening mammography for women aged 40–49. There is a new guideline synthesis, which includes only three of the guidelines—the US Preventive Services Task Force, the Kaiser Permanente, and the American College of Obstetricians and Gynecologists guidelines, the first two recommending against routine screening in women 40–49 and the third recommending for it. However, the American College of Obstetricians and Gynecologists guideline is a very mild positive recommendation—much more mild than the American Society of Breast Surgeons and the American College of Radiology recommendation—recommending only that women over 40 be "offered" screening mammography (i.e. not that they should definitely have screening mammography, which the American Society of Breast Surgeons and the American College of Radiology state). Moreover, the negative recommendations in the first three guidelines are framed as general recommendations that go together with a recommendation that individuals engage in joint decision making with their physicians. So, despite the updated guidelines and guideline synthesis, the situation remains the same: only some of the listed guidelines are used for the guideline synthesis; the guideline synthesis does not consider some of the major dissenters; and the guidelines that are included are framed to look as similar as possible. Note that this is my assessment of the situation.[21]

---

[21] I sent another inquiry to the National Guidelines Clearinghouse in July 2013 (which is when I first noticed the latest changes) to ask about the selection of guidelines for

When disagreement is "managed" in this way, it looks like there is more unanimity than there is. This serves to bolster the professional authority of medicine. I should add that I am not suggesting that such management of disagreement is deliberately deceptive, or that there is an explicit strategy of trying to maintain professional authority. It is likely that those who write the guideline syntheses have reasons for leaving out some of the guidelines (e.g. they may agree with Quanstrum and Hayward (2010) that radiologists have too much self-interest for their recommendations to be trusted). However, since there is not transparency about such reasons (and maybe it would be difficult or politically inadvisable to make all the reasons transparent), and the effect is to minimize disagreement, the result is management of disagreement and maintenance of authority.

There are 21 guideline syntheses listed on the National Guidelines Clearinghouse website. Yet there are over 2,500 guidelines listed, and my impression is that most common conditions have two to five listed guidelines that are relevant. Indeed, the National Guidelines Clearinghouse website has an online guideline comparison tool for use in comparing guidelines. For most conditions, there is not much controversy about clinical guidelines. What are the reasons for producing a guideline synthesis? The reasons are not stated and my inquiries about this have not provided clarity. My hypothesis is that guideline syntheses are produced for cases in which there is controversy. I am currently testing this hypothesis. (Other hypotheses are that guideline syntheses are produced for cases in which there are many guidelines for the same condition, or that guideline syntheses are produced for cases in which there is public interest.)

### 9.3.4  Structural Avoidance of Conflict

The case of screening mammography is not typical of clinical guidelines in two important respects. First, the evidence is "in the gray area" in the sense discussed above. Depending on how the data are analyzed, screening mammography may or may not be of overall benefit for women aged 40–49. For many recommended clinical interventions, the evidence is

guidelines synthesis. Vivian Coates, my contact person, gave me some criteria, but these criteria are not stated on the National Guidelines Clearinghouse website, and Vivian Coates says (in an e-mail to me dated June 24, 2013) that "we should probably make more transparent how we select guidelines for syntheses." I thank Vivian Coates for her responsiveness to my questions.

clearly positive, even if the intervention is less effective than we would like it to be. Second, consensus recommendations for screening mammography were produced *before* the era of evidence-based medicine and based on pathophysiological reasoning ("early detection saves lives") and clinical trials of poor quality. Belief perseverance phenomena, as well as professional interests, continue to pay a role. Since the end of the 1990s, it is general practice for there to be a systematic evidence review *before* a consensus conference on a recommendation,[22] making it less likely that the consensus conference approach will end up disagreeing with the evidence-based recommendations. The justification used for this is the philosophy of evidence-based medicine, which ranks consensus as the lowest level of evidence. This philosophy has demoted the role of consensus, but it has not removed it from the list of acceptable and even important epistemic methods.[23] The result (whatever the justification) is a structural means of generally avoiding conflict between evidence-based medicine and consensus conference approaches by giving priority to evidence-based medicine.

## 9.4  Managing Controversy in Medicine

Controversy about clinical recommendations is the exception rather than the rule in health care. Most (non-rare) diseases have standard practice guidelines. Physicians know that if they follow professional guidelines they can justify their actions as "the standard of care" to their peers and in legal contexts. To some degree, this standardization is the result of scientific consensus about effectiveness of interventions. To perhaps an equal degree, it is the result of structural means of avoiding controversy and active means of managing it.

The most common structural means of avoiding controversy, especially between evidence-based approaches and expert consensus, is to privilege evidence-based approaches by having them anchor expert assessments. This is accomplished by having evidence syntheses done (by one of AHRQ's evidence-based practice centers, or by a Cochrane review) before experts meet to put together clinical guidelines. For those

---

[22]  See Chapters 2 and 3 for details about this.
[23]  Recall the discussion in Section 6.6 about the evidence hierarchy itself, which is at crucial points justified by expert consensus.

who are fans of evidence-based medicine, and for those areas of medicine that are particularly amenable to producing evidence high on the hierarchy, this is an excellent way to proceed. Cystic fibrosis is an example of a disease in which there is minimal controversy about treatment despite continuing changes in treatment over time. Part of the reason for this is that the research and clinical communities are united in their commitment to the methods of evidence-based medicine.[24]

The general strategy of reframing small amounts and degrees of dissent as a locus for physician and patient autonomy is also widely followed.

## 9.5  Conclusions

A developing but untidy methodological pluralism is not a new situation in medicine. Since the time of Greek medicine, we have had empiric medicine, authority-based medicine, causal reasoning, and case-based reasoning, and while we have emphasized different methods at different times none has absolute hegemony over the others; none can be discarded; and each occasionally clashes with another. The same is true of the recent new methods in medicine: consensus conferences, evidence-based medicine, translational medicine, and narrative medicine. All the methods—indeed, all methods used in science, technology, and medicine—are indispensable, and all are fallible. There are no methodological shortcuts (such as general methodological hierarchies) to resolving controversy. However, there are general strategies for minimizing and avoiding public controversy. These should be balanced with an appreciation that expert disagreement can benefit both science and medicine.

---

[24] The cystic fibrosis community is unified in many ways that reduce the chance of significant dissent.

# 10

# Concluding Thoughts

This book has been a critical exploration of the epistemology of medicine, with a focus on the most recent developments. My priority has been to understand and represent, as charitably as possible, different methodologies and different views about those methodologies. I hope that my work is useful for understanding the epistemic characteristics of the different methods, and for seeing what is at stake in medical controversies. I aimed throughout for "epistemic equipoise," finding much of value in each of the methods selected for close examination. The traditional classification of methods into the "art" and the "science" of medicine has outlived its usefulness and I have replaced it with a developing, untidy, methodological pluralism.

Each method I have discussed has *something old* and *something new*. Authority-based medicine, which has always been a foundation for medical knowledge, is an ancestor of the medical consensus conference, which also aims to produce knowledge based on trust of authorities. The difference is that authority-based medicine rested on single authorities (say, Hippocrates or Galen), while consensus conference-based medicine makes use of the rhetorical power of the consensus of a group of experts. This rhetorical power is augmented by the ritual elements of consensus conferences (gravitas of proceedings, prestige of participants, openness of public portions, executive sessions for private deliberations, etc.), which are designed to maximize perceived objectivity. Evidence-based medicine has its roots in the long-established and worldwide practices of empiric medicine: treating symptoms with interventions that, through experience, have worked, whether or not there is an understanding of how the interventions work. Evidence-based medicine adds more rigorous methods of recording and aggregating experience. Translational medicine (T1, the main part of translational medicine) involves causal reasoning and informal experimentation, using hypotheses about the mechanisms involved and trial and error in interventions. The methods are traditional and somewhat inexplicit,[1] the hypotheses are up to the minute and fallible,[2] and the term "translational medicine" offers renewed hope that advances in basic research (such as the Human Genome Project) will yield practical benefits. Acknowledging and rewarding the early and messy stages of developing a useful intervention recognizes the kind of work that is necessary for the development of successful health care interventions. Finally, narrative medicine is rooted in at least two traditional elements: the use of case study, and the dyadic relationship between healer and patient. Instead of the traditional authoritarian healer–patient relationship, however, narrative medicine stresses a more contemporary egalitarian relationship between physician and patient.

Each method I have discussed has *something obvious* and *something odd*. It seems obvious that expert consensus should be taken seriously (one of our best arguments for the reality of anthropogenic climate change is that there is scientific consensus that it is occurring), yet strange that a scientific controversy is to be settled through discussion at a consensus conference—Trisha Greenhalgh registers her puzzlement by calling consensus conferences "GOBSAT," meaning "Good Old Boys Sat Around a Table" (2000: 6). It seems obvious that medicine is based on the evidence, yet perplexing that evidence-based medicine seems to dismiss a good deal of strong evidence. It is familiar that developing new health care interventions takes a good deal of trial and error, yet puzzling that the word "translation" is used in this context. It is a cliché that physicians should listen to what patients say, yet surprising that learning techniques of literary interpretation are helpful with listening in clinical contexts.

Each method I have discussed has an *ethos*: a set of associated values. For consensus conferences, the ethos includes earned authority (expertise), democracy, fairness, openness, and objectivity. For evidence-based medicine, the ethos includes analytic rigor, objectivity, and quantitative

---

[1] Unless Martin Wehling develops successful techniques for predicting which new interventions will work best, in which case the methods will become novel and more explicit. See the discussion of Wehling's ideas in Chapter 7.

[2] Currently we do not reason in terms of humors but in terms of genomics.

precision. For translational medicine the ethos includes fallibility, serendipity, persistence, interdisciplinarity, and creativity. And narrative medicine has an ethos of intimacy and ethicality.

Each method I have discussed has known *strengths* and *weaknesses*. Consensus conferences can help move a medical community to adopt new practices, but they can also fail to do so, either because the matter under discussion is quite controversial or because sufficient incentives to change practice habits are lacking. Evidence-based medicine techniques can help researchers avoid a variety of well-known biases, but they are not as reliable as expected. Translational medicine eschews inappropriate standards of rigor for the process of discovery, but often fails to produce therapeutic successes. Finally, narrative medicine can help provide humanistic care and do astonishing medical detective work, but can also lead the physician astray with information that has been misremembered, overlooked, or overemphasized in the effort to create a coherent narrative.

Each method I have discussed has *epistemic tensions* and *ironies*. I discussed these at length for consensus conferences, and they included tensions between expertise and intellectual bias, between scheduling too early and too late, between taking enough time to consider the issues in depth yet not so much time that busy professionals will refuse participation, being specific enough to have content yet not so specific that they dictate the practice of medicine, having a little controversy but not too much, having some process for review of the results but not too much, openness of some sessions, and secrecy of executive sessions. Evidence-based medicine has a tension between its demand for expertise (in statistical methods) and its claim to reject authority. There is tension between the demands of precise methodology and the need for "good judgment." It is also ironic that the details of the standard evidence hierarchy are to be settled by the consensus of experts in the GRADE Working Group. Narrative medicine shows tensions between the particularity of each case and general narrative forms, as well as between intimacy and professionalism and between identity and uniqueness.

"Personalized medicine" is a candidate for the next new methodology in medicine. "Personalized medicine" is a term that has been widely used since about 2000, primarily to describe therapies that are tailored to individual biochemistry such as the sequence of the somatic cell genome (or proteome) or the cancer cell genome. (The case of the CTL019 Protocol that saved Emma Whitehead, discussed in Chapter 7, is an example of personalized medicine as well as of translational medicine.) Its ethos is that it rejects the supposedly "one size fits all" and/or "cookbook" therapeutics of evidence-based medicine. This sounds at first like an ethos similar to that of narrative medicine, but the word "personalized" is a bit of a misnomer, since this meaning of personalized medicine has nothing to do with persons (taken in the humanistic sense). Perhaps confusion was facilitated by the pervasive cultural equation of our personhood with our DNA (Nelkin & Lindee 1995). In any case, the idea is that "personalized medicine" divides the patient population into subgroups based on differences in basic biochemistry. Such subgroups can be as small as groups of persons with the same genome (usually single individuals unless they are identical twins) or they can be larger, as in subgroups with the same cancer mutations.

Not surprisingly, the term "personalized medicine" has come to be used to refer to all kinds of personalization in medicine, and especially to a caring and personal physician–patient relationship. This is confused and confusing, although it has been rhetorically quite successful since it has enabled the marketing of "personalized medicine" to use a similar ethos (respect for individuality, intimacy, etc.) as narrative medicine. This ethos is attractive to many patients. In this way there is a unification of cutting-edge medical technologies (which might otherwise be quite scary) with personal care.

Is "personalized medicine" the next trend in epistemology of medicine? It has already achieved some prominence. I am not yet convinced that there is enough in the way of new methods, or even reclaimed old methods, to consider personalized medicine a new way of doing medical research and/or practice. Perhaps such methods will come in due course. Personalized medicine is rhetorically successful, bringing cutting-edge genomic science together with a rhetoric of individualized care, but it remains to be seen what substance lies behind the rhetoric.

Much of the time, the methods do not compete with each other; they each have roles to play, often at different stages of research. For example, translational medicine is appropriate for Phase I and Phase II clinical trials, while Phase III trials are the heart of evidence-based medicine. Medical institutions can be designed to minimize potential conflict between the methods (for example, by holding consensus conferences after formal evidence review). Some physicians make more use of narrative medicine techniques than do others—for example, those in primary care and psychiatry use them much more than specialists such as surgeons and radiologists. Nevertheless, as I argued in Chapter 9, the methods can produce results that conflict. I have described cases in which consensus judgment has disagreed with the results of evidence synthesis, cases in which medical narratives have conflicted with evidence-based medicine, and cases in which pathophysiological reasoning (in translational medicine) conflicts with the results of evidence-based medicine. In such cases, the strengths and weaknesses of each of the methods should be considered when exploring results further. There is no hierarchy of methods.

Methods in medicine are like solos in a jazz ensemble. They each have a turn in the limelight, and they build on earlier themes, sometimes fitting together, sometimes competing, and sometimes changing each other. None of the voices disappear, although they can fade into the background. No solo lasts forever. The music is always developing, never finished, and always a little rough around the edges.

# References

Ahrens, E.H. 1985, "The diet-heart question in 1985: has it really been settled?" *The Lancet*, vol. 325, no. 8437, pp. 1085–7.

AHRQ 2002, "Systems to rate the strength of scientific evidence," *Evidence Report/Technology Assessment 47* <http://archive.ahrq.gov/clinic/epcsums/strengthsum.htm> accessed October 30, 2014.

Als-Nielsen, B., Chen, W., Gluud, C., & Kjaergard, L.L. 2003, "Association of funding and conclusions in randomized drug trials: a reflection of treatment effect or adverse events?" *JAMA: The Journal of the American Medical Association*, vol. 290, no. 7, pp. 921–8.

Andersen, H. 2012, "Mechanisms: what are they evidence for in evidence-based medicine?" *Journal of Evaluation in Clinical Practice*, vol. 18, no. 5, pp. 992–9.

Andersen, I. & Jaeger, B. 1999, "Scenario workshops and consensus conferences: towards more democratic decision-making," *Science and Public Policy*, vol. 26, no. 5, pp. 331–40.

Anderson, E. 2004, "Uses of value judgments in science: a general argument, with lessons from a case study of feminist research on divorce," *Hypatia*, vol. 19, no. 1, pp. 1–24.

Andreasen, P.B. 1988, "Consensus conferences in different countries: aims and perspectives," *International Journal of Technology Assessment in Health Care*, vol. 4, pp. 305–8.

Angell, M. 2005, *The truth about the drug companies: how they deceive us and what to do about it*, Revised and updated edition, Random House Trade Paperbacks, New York.

Anglemyer, A., Horvath, H.T., & Bero, L. 2014, "Healthcare outcomes assessed with observational study designs compared with those assessed in randomized trials," *The Cochrane Database of Systematic Reviews*, vol. 4, p. MR000034.

Asch, S.J. & Lowe, C.U. 1984, "The Consensus Development Program: theory, process and critique," *Knowledge: Creation, Diffusion, Utilization*, vol. 5, no. 3, pp. 369–85.

Ashcroft, R.E. 2004, "Current epistemological problems in evidence based medicine," *Journal of Medical Ethics*, vol. 30, no. 2, pp. 131–5.

Bachelard, G. 1953, *Le materialisme rationnel*, Presses Universitaires de France, Paris.

Backer, B. 1990, "Profile of the Consensus Development Program in Norway: The Norwegian Institute for Hospital Research and the National Research

Council," in *Improving consensus development for health technology assessment: an international perspective*, eds. C. Goodman & S.R. Baratz, National Academy Press, Washington, DC, pp. 118–24.

Barad, K.M. 2007, *Meeting the universe halfway: quantum physics and the entanglement of matter and meaning*, Duke University Press, Durham, NC.

Battista, R.N. 1990, "Profile of a Consensus Development Program in Canada: The Canadian Task Force on the Periodic Health Examination," in *Improving consensus development for health technology assessment: an international perspective*, eds. C. Goodman & S.R. Baratz, National Academy Press, Washington, DC, pp. 87–95.

Baum, M. 2004, "Commentary: false premises, false promises and false positives—the case against mammographic screening for breast cancer," *International Journal of Epidemiology*, vol. 33, no. 1, pp. 66–7; discussion 69–73.

Baum, M., Demicheli, R., Hrushesky, W., & Retsky, M. 2005, "Does surgery unfavourably perturb the 'natural history' of early breast cancer by accelerating the appearance of distant metastases?" *European Journal of Cancer*, vol. 41, pp. 508–15.

Bechtel, W. & Richardson, R.C. 1993, *Discovering complexity: decomposition and localization as strategies in scientific research*, Princeton University Press, Princeton, NJ.

Bekelman, J.E., Li, Y., & Gross, C.P. 2003, "Scope and impact of financial conflicts of interest in biomedical research: a systematic review," *JAMA: The Journal of the American Medical Association*, vol. 289, no. 4, pp. 454–65.

Benson, K. & Hartz, A.J. 2000, "A comparison of observational studies and randomized, controlled trials," *New England Journal of Medicine*, vol. 342, no. 25, pp. 1878–86.

Bernard, C. 1957, *An introduction to the study of experimental medicine*, Dover Publications, New York.

Berwick, D.M. 2005, "Broadening the view of evidence-based medicine," *Quality & Safety in Health Care*, vol. 14, no. 5, pp. 315–16.

Bird, A. 2011, "What can philosophy tell us about evidence-based medicine? An assessment of Jeremy Howick's *The philosophy of evidence-based medicine*," *The International Journal of Person Centered Medicine*, vol. 1, no. 4, pp. 642–8.

Bliss, M. 1982, *The discovery of insulin*, University of Chicago Press, Chicago.

Bluhm, R. 2011, "Jeremy Howick: the philosophy of evidence-based medicine," *Theoretical Medicine and Bioethics*, vol. 32, pp. 423–7.

Bluhm, R. 2005, "From hierachy to network: a richer view of evidence for evidence-based medicine," *Perspectives in Biology and Medicine*, vol. 48, no. 4, pp. 535–47.

Bluhm, R. & Borgerson, K. 2011, "Evidence-based medicine," in *Philosophy of Medicine*, ed. F. Gifford, Elsevier, Amsterdam, pp. 203–38.

Boden, M.A. 2004, *The creative mind: myths and mechanisms*, 2nd edition, Routledge, London; New York.

Boden, M.A. 1994, *Dimensions of creativity*, MIT Press, Cambridge, MA.

Bohlin, I. 2012, "Formalizing syntheses of medical knowledge: the rise of meta-analysis and systematic reviews," *Perspectives on Science*, vol. 20, no. 3, pp. 273–309.

Braude, H.D. 2012, *Intuition in medicine: a philosophical defense of clinical reasoning*, University of Chicago Press, Chicago.

Brody, H. 2003 [1987], *Stories of sickness*, 2nd edition, Oxford University Press, Oxford; New York.

Brozek, J.L., Akl, E.A., Alonso-Coello, P., Lang, D., Jaeschke, R., Williams, J.W., Phillips, B., Lelgemann, M., Lethaby, A., Bousquet, J., Guyatt, G.H., Schunemann, H.J., & GRADE Working Group 2009a, "Grading quality of evidence and strength of recommendations in clinical practice guidelines: Part 1 of 3. An overview of the GRADE approach and grading quality of evidence about interventions," *Allergy*, vol. 64, no. 5, pp. 669–77.

Brozek, J.L., Akl, E.A., Jaeschke, R., Lang, D.M., Bossuyt, P., Glasziou, P., Helfand, M., Ueffing, E., Alonso-Coello, P., Meerpohl, J., Phillips, B., Horvath, A.R., Bousquet, J., Guyatt, G.H., Schunemann, H.J. & GRADE Working Group 2009b, "Grading quality of evidence and strength of recommendations in clinical practice guidelines: Part 2 of 3. The GRADE approach to grading quality of evidence about diagnostic tests and strategies," *Allergy*, vol. 64, no. 8, pp. 1109–16.

Bruner, J. 1986, *Actual minds, possible worlds*, Harvard University Press, Cambridge, MA.

Buchbinder, R., Osborne, R.H., Ebeling, P.R., Wark, J.D., Mitchell, P., Wriedt, C., Graves, S., Staples, M.P., & Murphy, B. 2009, "A randomized trial of vertebroplasty for painful osteoporotic vertebral fractures," *New England Journal of Medicine*, vol. 361, no. 6, pp. 557–68.

Butler, D. 2008, "Translational research: crossing the valley of death," *Nature*, vol. 453, no. 7197, pp. 840–2.

Calltorp, J. 1988, "Consensus development conferences in Sweden: effects on health policy and administration," *International Journal of Technology Assessment in Health Care*, vol. 4, pp. 75–88.

Carel, H. 2012, "Phenomenology as a resource for patients," *The Journal of Medicine and Philosophy*, vol. 37, no. 2, pp. 96–113.

Carel, H. 2008, *Illness: the cry of the flesh*, Acumen, Stocksfield.

Cartwright, N. 2012, "Will this policy work for you? Predicting effectiveness better: how philosophy helps," *Philosophy of Science*, vol. 79, no. 5, pp. 973–89.

Cartwright, N. 2009, "Evidence-based policy: what's to be done about relevance," *Philosophical Studies*, vol. 143, no. 1, pp. 127–36.

Cartwright, N. 2007a, "Are RCTs the Gold Standard?" *Biosocieties*, vol. 2, no. 2, pp. 11–20.

Cartwright, N. 2007b, "Evidence-based policy: where is our theory of evidence?" *Center for Philosophy of Natural and Social Science, London School of Economics, Technical Report 07/07*.

Cartwright, N. 1999, *The dappled world: a study of the boundaries of science*, Cambridge University Press, Cambridge; New York.

Cartwright, N. 1989, *Nature's capacities and their measurement*, Clarendon Press; Oxford University Press, Oxford; New York.

Cartwright, N. 1983, *How the laws of physics lie*, Clarendon Press; Oxford University Press, Oxford; New York.

Cartwright, N. & Munro, E. 2010, "The limitations of randomized controlled trials in predicting effectiveness," *Journal of Evaluation in Clinical Practice*, vol. 16, no. 2, pp. 260–6.

Cassell, E.J. 1991, *The nature of suffering and the goals of medicine*, Oxford University Press, New York.

Cassell, E.J. 1976, *The healer's art: a new approach to the doctor–patient relationship*, 1st edition, Lippincott, Philadelphia.

Chabner, B.A., Boral, A.L., & Multani, P. 1998, "Translational research: walking the bridge between idea and cure—seventeenth Bruce F. Cain Memorial Award lecture," *Cancer Research*, vol. 58, no. 19, pp. 4211–16.

Chalmers, I., Enkin, M., & Keirse, M.J.N.C. 1989, *Effective care in pregnancy and childbirth*, Oxford Medical Publications, Oxford; New York.

Chang, H. 2004, *Inventing temperature: measurement and scientific progress*, Oxford University Press, New York.

Charlton, B.G. & Miles, A. 1998, "The rise and fall of EBM," *Quarterly Journal of Medicine*, vol. 91, no. 5, pp. 371–4.

Charon, R. 2006a, *Narrative medicine: honoring the stories of illness*, Oxford University Press, Oxford; New York.

Charon, R. 2006b, "The self-telling body," *Narrative Inquiry*, vol. 16, no. 1, pp. 191–200.

Charon, R. 2004, "Narrative and medicine," *New England Journal of Medicine*, vol. 350, no. 9, pp. 862–4.

Charon, R., Wyer, P., & NEBM Working Group 2008, "Narrative evidence based medicine," *Lancet*, vol. 371, no. 9609, pp. 296–7.

Chintu, C., Bhat, G.J., Walker, A.S., Mulenga, V., Sinyinza, F., Lishimpi, K., Farrelly, L., Kaganson, N., Zumla, A., Gillespie, S.H., Nunn, A.J., Gibb, D.M., & CHAP trial team 2004, "Co-trimoxazole as prophylaxis against opportunistic infections in HIV-infected Zambian children (CHAP): a double-blind randomised placebo-controlled trial," *Lancet*, vol. 364, no. 9448, pp. 1865–71.

Chisholm, R.M. 1989, *Theory of knowledge*, 3rd edition, Prentice-Hall International, Englewood Cliffs, NJ; London.

Clarke, B., Gillies, D., Illari, P., Russo, F., & Williamson, J. 2013, "The evidence that evidence-based medicine omits," *Preventive Medicine*, vol. 5, no. 6, pp. 745–7.

Clough, S. 2004, "Having it all: naturalized normativity in feminist science studies," *Hypatia*, vol. 19, no. 1, pp. 102–18.

Cohen, A.M., Stavri, P.S., & Hersh, W.R. 2004, "A categorization and analysis of the criticisms of evidence-based medicine," *International Journal of Medical Informatics*, vol. 73, no. 1, pp. 35–43.

Collins, H.M. 1985, *Changing order: replication and induction in scientific practice*, Sage Publications, London; Beverly Hills, CA.

Collins, H.M. & Evans, R. 2007, *Rethinking expertise*, University of Chicago Press, Chicago.

Collins, H.M. & Pinch, T.J. 1993, *The golem: what everyone should know about science*, Cambridge University Press, Cambridge; New York.

Concato, J., Shah, N., & Horwitz, R.I. 2000, "Randomized, controlled trials, observational studies, and the hierarchy of research designs," *New England Journal of Medicine*, vol. 342, no. 25, pp. 1887–92.

Contopoulos-Ioannidis, D.G., Ntzani, E., & Ioannidis, J.P. 2003, "Translation of highly promising basic science research into clinical applications," *The American Journal of Medicine*, vol. 114, no. 6, pp. 477–84.

Coulter, I., Adams, A., & Shekelle, P. 1995, "Impact of varying panel membership on ratings of appropriateness in consensus panels: a comparison of a multi-and single disciplinary panel," *Health Services Research*, vol. 30, no. 4, pp. 577–91.

Coutsoudis, A., Pillay, K., Spooner, E., Coovadia, H.M., Pembrey, L., & Newell, M.L. 2005, "Routinely available cotrimoxazole prophylaxis and occurrence of respiratory and diarrhoeal morbidity in infants born to HIV-infected mothers in South Africa," *South African Medical Journal [Suid-Afrikaanse tydskrif vir geneeskunde]*, vol. 95, no. 5, pp. 339–45.

Daly, J. 2005, *Evidence-based medicine and the search for a science of clinical care*, University of California Press; Milbank Memorial Fund, Berkeley, CA; New York.

Darden, L. 2010, *Mechanisms, mutations and rational drug therapy in the case of cystic fibrosis*, unpublished talk given at the History of Science Society meetings in 2010.

Darden, L. 2006, *Reasoning in biological discoveries: essays on mechanisms, interfield relations, and anomaly resolution*, Cambridge University Press, New York.

Davidoff, F., Haynes, B., Sackett, D., & Smith, R. 1995, "Evidence based medicine," *British Medical Journal (Clinical Research Ed.)*, vol. 310, no. 6987, pp. 1085–6.

Delborne, J.A., Anderson, A.A., Kleinman, D.L., Colin, M., & Powell, M. 2009, "Virtual deliberation? Prospects and challenges for integrating the Internet in consensus conferences," *Public Understanding of Science*, vol. 20, no. 3, pp. 367–84.

Douglas, H.E. 2009, *Science, policy, and the value-free ideal*, University of Pittsburgh Press, Pittsburgh, PA.

Dreger, A. 2000, "Metaphors of morality in the Human Genome Project," in *Controlling our destinies: historical, philosophical, ethical, and theological perspectives on the Human Genome Project*, ed. P. Sloan, University of Notre Dame Press, IN, pp. 155–84.

Dressler, L. 2006, *Consensus through conversation: how to achieve high-commitment decisions*, Berrett-Koehler, San Francisco, CA.

Dupré, J. 1993, *The disorder of things: metaphysical foundations of the disunity of science*, Harvard University Press, Cambridge, MA.

Eddy, D.M. 2005, "Evidence-based medicine: a unified approach," *Health Affairs (Project Hope)*, vol. 24, no. 1, pp. 9–17.

Ehrenreich, B. 2001, "Welcome to cancerland," *Harper's Magazine*, November 2001, pp. 43–53.

Elstein, A.S. 2004, "On the origins and development of evidence-based medicine and medical decision making," *Inflammation Research*, vol. 53, Supplement 2, pp. S184–S189.

Elstein, A.S. 1999, "Heuristics and biases: selected errors in clinical reasoning," *Academic Medicine: Journal of the Association of American Medical Colleges*, vol. 74, no. 7, pp. 791–4.

Engel, G.L. 1977, "The need for a new medical model," *Science*, vol. 196, no. 4286, pp. 129–36.

Epstein, S. 1996, *Impure science: AIDS, activism, and the politics of knowledge*, University of California Press, Berkeley.

Ericksen, J.A. 2008, *Taking charge of breast cancer*, University of California Press, Berkeley.

Erikson, K.T. 1976, *Everything in its path: destruction of community in the Buffalo Creek Flood*, Simon and Schuster, New York.

Eva, K.W. & Norman, G.R. 2005, "Heuristics and biases: a biased perspective on clinical reasoning," *Medical Education*, vol. 39, no. 9, pp. 870–2.

Evidence-Based Medicine Working Group 1992, "Evidence-based medicine: A new approach to teaching the practice of medicine," *JAMA: The Journal of the American Medical Association*, vol. 268, no. 17, pp. 2420–5.

Falkum, E. 2008, "Phronesis and techne: the debate on evidence-based medicine in psychiatry and psychotherapy," *Philosophy, Psychiatry, and Psychology*, vol. 15, no. 2, pp. 141–9.

Ferguson, J.H. 1999, "Curative and population medicine: bridging the great divide," *Neuroepidemiology*, vol. 18, no. 3, pp. 111–19.

Ferguson, J.H. 1997, "Interpreting scientific evidence: comparing the National Institutes of Health Consensus Development Program and courts of law," *The Judges' Journal*, Summer, pp. 21–4; 83–4.

Ferguson, J.H. 1995, "The NIH consensus development program," *The Joint Commission Journal on Quality Improvement*, vol. 21, no. 7, pp. 332–6.

Ferguson, J.H. 1993, "NIH consensus conferences: dissemination and impact," *Annals of the New York Academy of Sciences*, vol. 703, pp. 180–98; discussion 198–9.

Ferguson, J.H. & Sherman, C.R. 2001, "Panelists' views of 68 NIH Consensus Conference," *International Journal of Technology Assessment in Health Care*, vol. 17, no. 4, pp. 542–58.

Fergusson, D., Glass, K.C., Waring, D., & Shapiro, S. 2004, "Turning a blind eye: the success of blinding reported in a random sample of randomised, placebo controlled trials," *BMJ (Clinical Research Ed.)*, vol. 328, no. 7437, p. 432.

Fine, A. 1986, *The shaky game: Einstein, realism, and the quantum theory*, University of Chicago Press, Chicago.

Fink, A., Kosecoff, J., Chassin, M., & Brook, R.H. 1984, "Consensus methods: characteristics and guidelines for use," *American Journal of Public Health*, vol. 74, no. 9, pp. 979–83.

Flume, P.A., O'Sullivan, B.P., Robinson, K.A., Goss, C.H., Mogayzel, P.J., Jr., Willey-Courand, D.B., Bujan, J., Finder, J., Lester, M., Quittell, L., Rosenblatt, R., Vender, R.L., Hazle, L., Sabadosa, K., Marshall, B. & Cystic Fibrosis Foundation, Pulmonary Therapies Committee 2007, "Cystic fibrosis pulmonary guidelines: chronic medications for maintenance of lung health," *American Journal of Respiratory and Critical Care Medicine*, vol. 176, no. 10, pp. 957–69.

Frank, A.W. 1995, *The wounded storyteller: body, illness, and ethics*, University of Chicago Press, Chicago.

Fraser, N. 1990, "Rethinking the public sphere: a contribution to the critique of actually existing democracy," *Social Text*, vol. 25/26, pp. 56–80.

Fredrickson, D.S. 1978, "Seeking technical consensus on medical interventions," *Clinical Research*, pp. 116–17.

Freidson, E. 2001, *Professionalism: the third logic*, University of Chicago Press, Chicago.

Fricker, M. 2007, *Epistemic injustice: power and the ethics of knowing*, Oxford University Press, Oxford; New York.

Galison, P.L. 1997, *Image and logic: a material culture of microphysics*, University of Chicago Press, Chicago.

Galison, P.L. 1987, *How experiments end*, University of Chicago Press, Chicago.

Gallese, V. & Goldman, A. 1998, "Mirror neurons and the simulation theory of mind-reading," *Trends in Cognitive Science*, vol. 2, no. 12, pp. 493–501.

Gawande, A. 2013, "How do good ideas spread?" *The New Yorker*, July 29, 2013.

Gawande, A. 2004, "The Bell Curve," *The New Yorker*, December 6, 2004.

Gelderloos, P. 2006, *Consensus: a new handbook for grassroots social, political, and environmental groups*, See Sharp Press, Tucson, AZ.

Giere, R.N. 2006, *Scientific perspectivism*, University of Chicago Press, Chicago.

Giere, R.N. 1999, *Science without laws*, University of Chicago Press, Chicago.

Giere, R.N. 1988, *Explaining science: a cognitive approach*, University of Chicago Press, Chicago.

Gifford, F. 2011, *Philosophy of medicine*, 1st edition, North Holland, Oxford; Burlingon, MA.

Glasziou, P., Chalmers, I., Rawlins, M., & McCulloch, P. 2007, "When are randomised trials unnecessary? Picking signal from noise," *BMJ (Clinical Research Ed.)*, vol. 334, no. 7589, pp. 349–51.

Goldman, A.I. 2006, *Simulating minds: the philosophy, psychology, and neuroscience of mindreading*, Oxford University Press, Oxford; New York.

Goldman, A.I. 2002, *Pathways to knowledge: private and public*, Oxford University Press, Oxford; New York.

Goldman, A.I. 1999, *Knowledge in a social world*, Clarendon Press; Oxford University Press, Oxford; New York.

Goldman, A.I. 1992, *Liaisons: philosophy meets the cognitive and social sciences*, MIT Press, Cambridge, MA.

Goodman, K.W. 2003, *Ethics and evidence-based medicine: fallibility and responsibility in clinical science*, Cambridge University Press, New York.

Goodman, N. 1955, *Fact, fiction, and forecast*, Harvard University Press, Cambridge.

Goodman, S. 2008, "A dirty dozen: twelve p-value misconceptions," *Seminars in Hematology*, vol. 45, no. 3, pp. 135–40.

Goodman, S. 2002, "The mammography dilemma: a crisis for evidence-based medicine?" *Annals of Internal Medicine*, vol. 137, no. 5, Part 1, pp. 363–5.

Gotzsche, P.C. & Nielsen, M. 2011, "Screening for breast cancer with mammography," *Cochrane Database of Systematic Reviews (Online)*, vol. 1, no. 1, p. CD001877.

Gotzsche, P.C. & Olsen, O. 2000, "Is screening for breast cancer with mammography justifiable?" *Lancet*, vol. 355, no. 9198, pp. 129–34.

Gould, S.J. 1985, "The median isn't the message," *Discover*, vol. 6, June, pp. 40–2.

Goyal, R.K., Charon, R., Lekas, H.M., Fullilove, M.T., Devlin, M.J., Falzon, L., & Wyer, P.C. 2008, "'A local habitation and a name': how narrative evidence-based medicine transforms the translational research paradigm," *Journal of Evaluation in Clinical Practice*, vol. 14, no. 5, pp. 732–41.

Greenhalgh, T. 2000, *How to read a paper*, 2nd edition, BMJ, London.

Greenhalgh, T. & Nuffield Trust for Research and Policy Studies in Health Services 2006, *What seems to be the trouble? Stories in illness and healthcare*, Radcliffe, Oxford; Seattle.

Greenhalgh, T., Wong, G., Westhorp, G., & Pawson, R. 2011, "Protocol—realist and meta-narrative evidence synthesis: evolving standards (RAMESES)," *BMC Medical Research Methodology*, vol. 11, p. 115.

Greer, A.L. 1987, "The two cultures of biomedicine: can there be consensus?" *JAMA: The Journal of the American Medical Association*, vol. 258, no. 19, pp. 2739–40.

Groopman, J.E. 2008, *How doctors think*, Mariner Books edition, Houghton Mifflin, Boston.

Grossman, J. & Mackenzie, F. 2005, "The randomized controlled trial: Gold Standard, or merely standard?" *Perspectives in Biology and Medicine*, vol. 48, no. 4, pp. 516–34.

Gunby, P. 1980, "Pros and cons of NIH consensus conferences," *JAMA: The Journal of the American Medical Association*, vol. 244, no. 13, pp. 1413–14.

Guston, D. 1999, "Evaluating the first US consensus conference: the impact of the citizens' panel on telecommunications and the future of democracy," *Science, Technology and Human Values*, vol. 24, no. 4, pp. 451–82.

Guyatt, G.H., Oxman, A.D., Vist, G.E., Kunz, R., Falck-Ytter, Y., Alonso-Coello, P., Schunemann, H.J., & GRADE Working Group 2008, "GRADE: an emerging consensus on rating quality of evidence and strength of recommendations," *BMJ (Clinical Research Ed.)*, vol. 336, no. 7650, pp. 924–6.

Habermas, J. 1971, *Knowledge and human interests*, Beacon Press, Boston.

Hacking, I. 1983, *Representing and intervening: introductory topics in the philosophy of natural science*, Cambridge University Press, Cambridge; New York.

Halpern, J. 2001, *From detached concern to empathy: humanizing medical practice*, Oxford University Press, Oxford; New York.

Hanlon, C.R. 1981, "Letter to the editor," *CA: A Cancer Journal for Clinicians*, vol. 1, no. 31, p. 56.

Hanson, N.R. 1958, *Patterns of discovery: an inquiry into the conceptual foundations of science*, Cambridge University Press, Cambridge.

Harari, E. 2001, "Whose evidence? Lessons from the philosophy of science and the epistemology of medicine," *Australian and New Zealand Journal of Psychiatry*, vol. 35, no. 6, pp. 724–30.

Haraway, D.J. 1991, *Simians, cyborgs, and women: the reinvention of nature*, Routledge, New York.

Haraway, D.J. 1989, *Primate visions: gender, race, and nature in the world of modern science*, Routledge, New York.

Harding, S. 1993, "Rethinking standpoint epistemology: what is 'strong objectivity'?" in *Feminist Epistemologies*, eds. L. Alcoff & E. Potter, Routledge, New York and London, pp. 49–82.

Harding, S.G. 1991, *Whose science? Whose knowledge? Thinking from women's lives*, Cornell University Press, Ithaca, NY.

Hardwig, J. 1997, "Autobiography, biography, and narrative ethics," in *Stories and their limits: narrative approaches to bioethics*, ed. H.L. Nelson, Routledge, New York, pp. 50–64.

Hardwig, J. 1991, "The role of trust in knowledge," *The Journal of Philosophy*, vol. 88, no. 12, pp. 693–708.

Hess, H.H. 1962, "History of ocean basins," in *Petrologic Studies: A Volume to Honor A.F. Buddington*, eds. A.E.J. Engel, H.L. James, & B.F. Leonard, Geological Society of America, New York, pp. 599–620.

Hesse, M.B. 1966, *Models and analogies in science*, University of Notre Dame Press, Notre Dame, IN.

Hill, A.B. 1952, "The clinical trial," *New England Journal of Medicine*, vol. 247, no. 4, pp. 113–19.

Hill, M.N. 1991, "Physicians' perceptions of consensus reports," *International Journal of Technology Assessment in Health Care*, vol. 7, no. 1, pp. 30–41.

Holleb, A.I. 1980, "Medicine by Proxy?" *CA: A Cancer Journal for Clinicians*, vol. 30, no. 3, pp. 191–2.

Hopewell, S., Clarke, M.J., Stewart, L., & Tierney, J. 2007, "Time to publication for results of clinical trials," *Cochrane Database of Methodology Reviews*, issue 2, no. MR000011.

Horig, H. & Pullman, W. 2004, "From bench to clinic and back: perspective on the 1st IQPC Translational Research Conference," *Journal of Translational Medicine*, vol. 2, no. 1, p. 44.

Horst, M. & Irwin, A. 2010, "Nations at ease with radical knowledge: on consensus, consensusing and false consensusness," *Social Studies of Science*, vol. 40, no. 1, pp. 105–26.

Howick, J. 2011a, "Exposing the vanities—and a qualified defense—of mechanistic reasoning in health care decision making," *Philosophy of Science*, vol. 78, no. 5, pp. 926–40.

Howick, J. 2011b, *The philosophy of evidence-based medicine*, Wiley-Blackwell, Oxford.

Howick, J. 2008, "Against *a priori* judgments of bad methodology: questioning double-blinding as a universal methodological virtue of clinical trials," Philosophy of Science Association 2008 contributed conference paper, archived at <http://philsci-archive.pitt.edu/4279/> accessed October 30, 2014.

Hull, D.L. 1988, *Science as a process: an evolutionary account of the social and conceptual development of science*, University of Chicago Press, Chicago.

Iacoboni, M., Molnar-Szakacs, I., Gallese, V., Buccino, G., Mazziotta, J.C., & Rizzolatti, G. 2005, "Grasping the intentions of others with one's own mirror neuron system," *PLoS Biology*, vol. 3, no. 3, p. e79.

Institute of Medicine 1990a, *Consensus development at NIH: improving the program*, National Academy Press, Washington, DC.

Institute of Medicine 1990b, *Improving consensus development for health technology assessment: an international perspective*, National Academy Press, Washington, DC.

Institute of Medicine 1985, *Assessing medical technologies*, National Academy Press, Washington, DC.

Ioannidis, J.P. 2005, "Contradicted and initially stronger effects in highly cited clinical research," *JAMA: The Journal of the American Medical Association*, vol. 294, no. 2, pp. 218–28.

Ioannidis, J.P., Haidich, A.B., Pappa, M., Pantazis, N., Kokori, S.I., Tektonidou, M.G., Contopoulos-Ioannidis, D.G., & Lau, J. 2001, "Comparison of evidence of treatment effects in randomized and nonrandomized studies," *JAMA: The Journal of the American Medical Association*, vol. 286, no. 7, pp. 821–30.

Jacoby, I. 1990, "Sponsorship and role of consensus development programs within national health care systems," in *Improving consensus development for health technology assessment: an international perspective*, eds. C. Goodman & S. Baratz, National Academy Press, Washington, DC, pp. 7–17.

Jacoby, I. 1988, "Evidence and consensus," *JAMA: The Journal of the American Medical Association*, vol. 259, no. 20, p. 3039.

Jacoby, I. 1985, "The consensus development program of the National Institutes of Health: current practices and historical perspectives," *International Journal of Technology Assessment in Health Care*, vol. 1, no. 2, pp. 420–32.

Jacoby, I. 1983, "Biomedical technology: information dissemination and the NIH consensus development process," *Knowledge: Creation, Diffusion, Utilization*, vol. 5, no. 2, pp. 245–61.

Jacoby, I. & Clark, S.M. 1986, "Direct mailing as a means of disseminating NIH consensus conferences," *JAMA: The Journal of the American Medical Association*, vol. 255, no. 10, pp. 1328–30.

Janis, I.L. 1982, *Groupthink: psychological studies of policy decisions and fiascoes*, 2nd edition, Houghton Mifflin, Boston.

Janis, I.L. 1972, *Victims of groupthink: a psychological study of foreign-policy decisions and fiascoes*, Houghton Mifflin, Oxford.

Jasanoff, S. 2005, *Designs on nature: science and democracy in Europe and the United States*, Princeton University Press, Princeton, NJ.

Jasanoff, S. 2004, *States of knowledge: the co-production of science and social order*, Routledge, London; New York.

Johnsson, M. 1988, "Evaluation of the consensus conference program in Sweden: its impact on physicians," *International Journal of Technology Assessment in Health Care*, vol. 4, pp. 89–94.

Jorgensen, T. 1990, "Profile of the Consensus Development Program in Denmark: The Danish Medical Research Council and the Danish Hospital Institute," in *Improving consensus development for health technology assessment: an international perspective*, eds. C. Goodman & S.R. Baratz, National Academy Press, Washington, DC, pp. 96–101.

Juni, P., Witschi, A., Bloch, R., & Egger, M. 1999, "The hazards of scoring the quality of clinical trials for meta-analysis," *JAMA: The Journal of the American Medical Association*, vol. 282, no. 11, pp. 1054–60.

Kahan, J.P., Kanouse, D.E., & Winkler, J.D. 1988, "Stylistic variations in National Institutes of Health consensus statements, 1979–1983," *International Journal of Technology Assessment in Health Care*, vol. 4, no. 2, pp. 289–304.

Kahan, J.P., Kanouse, D.E., & Winkler, J.D. 1984, *Variations in the content and style of NIH consensus statements 1979–83*, Rand Corporation, Santa Monica, CA.

Kahneman, D., Slovic, P., & Tversky, A. 1982, *Judgments under uncertainty: heuristics and biases*, Cambridge University Press, Cambridge.

Kalberer, J.T. 1985, "Peer review and the consensus development process," *Science, Technology and Human Values*, vol. 10, no. 3, pp. 63–72.

Kallmes, D.F., Comstock, B.A., Heagerty, P.J., Turner, J.A., Wilson, D.J., Diamond, T.H., Edwards, R., Gray, L.A., Stout, L., Owen, S., Hollingworth, W., Ghdoke, B., Annesley-Williams, D.J., Ralston, S.H., & Jarvik, J.G. 2009, "A randomized trial of vertebroplasty for osteoporotic spinal fractures," *New England Journal of Medicine*, vol. 361, no. 6, pp. 569–79.

Kantrowitz, A. 1967, "Proposal for an institution for scientific judgment," *Science*, vol. 156, pp. 763–4.

Kantrowitz, A., Kennedy, D., & Seitz, F. 1976, "The Science Court Experiment: an interim report," *Science*, vol. 193, no. 4254, pp. 653–6.

Katz, J. 1984, *The silent world of doctor and patient*, Free Press, London; Collier Macmillan, New York.

Keller, E.F. 1985, *Reflections on gender and science*, Yale University Press, New Haven, CT.

Keller, E.F. 1983, *A feeling for the organism: the life and work of Barbara McClintock*, W.H. Freeman, San Francisco.

Kerner, J.F. 2006, "Knowledge translation versus knowledge integration: a 'funder's' perspective," *The Journal of Continuing Education in the Health Professions*, vol. 26, no. 1, pp. 72–80.

Kevles, D.J. & Hood, L.E. 1992, *The code of codes: scientific and social issues in the Human Genome Project*, Harvard University Press, Cambridge, MA.

Khoury, M.J., Gwinn, M., Yoon, P.W., Dowling, N., Moore, C.A., & Bradley, L. 2007, "The continuum of translation research in genomic medicine: how can we accelerate the appropriate integration of human genome discoveries into health care and disease prevention?" *Genetics in Medicine: Official Journal of the American College of Medical Genetics*, vol. 9, no. 10, pp. 665–74.

Kirkley, A., Birmingham, T.B., Litchfield, R.B., Giffin, J.R., Willits, K.R., Wong, C.J., Feagan, B.G., Donner, A., Griffin, S.H., D'Ascanio, L.M., Pope, J.E., & Fowler, P.J. 2008, "A randomized trial of arthroscopic surgery for osteoarthritis of the knee," *New England Journal of Medicine*, vol. 359, no. 11, pp. 1097–107.

Kitcher, P. 2001, *Science, truth, and democracy*, Oxford University Press, Oxford; New York.

Kitcher, P. 1993, *The advancement of science: science without legend, objectivity without illusions*, Oxford University Press, New York.

Klein, J.G. 2005, "Five pitfalls in decisions about diagnosis and prescribing," *BMJ (Clinical Research Ed.)*, vol. 330, no. 7494, pp. 781–3.

Kleinman, A. 1988, *The illness narratives: suffering, healing, and the human condition*, Basic Books, New York.

Kola, I. & Landis, J. 2004, "Opinion: can the pharmaceutical industry reduce attrition rates?" *Nature Reviews Drug Discovery*, vol. 3, no. 8, pp. 711–16.

Kolata, G. 1985, "Heart panel's conclusions questioned," *Science*, vol. 277, no. 4682, pp. 40–1.

Kosecoff, J., Kanouse, D.E., & Brook, R.H. 1990, "Changing practice patterns in the management of primary breast cancer: Consensus Development Program," *Health Services Research*, vol. 25, no. 5, pp. 809–23.

Kosecoff, J., Kanouse, D.E., Rogers, W.H., McCloskey, L., Winslow, C.M., & Brook, R.H. 1987, "Effects of the National Institutes of Health Consensus Development Program on physician practice," *JAMA: The Journal of the American Medical Association*, vol. 258, no. 19, pp. 2708–13.

Krimsky, S. 2003, *Science in the private interest: has the lure of profits corrupted biomedical research?* Rowman & Littlefield, Lanham, MD.

Kuhn, T.S. 1977, *The essential tension: selected studies in scientific tradition and change*, University of Chicago Press, Chicago.

Kuhn, T.S. 1962, *The structure of scientific revolutions*, University of Chicago Press, Chicago.

Kukla, R. 2005, *Mass hysteria: medicine, culture, and mothers' bodies*, Rowman & Littlefield, Lanham, MD.

Kunz, R., Burnand, B., Schunemann, H.J., & Grading of Recommendations, Assessment, Development and Evaluation (GRADE) Working Group 2008, "The GRADE System: an international approach to standardize the graduation of evidence and recommendations in guidelines," *Der Internist*, vol. 49, no. 6, pp. 673–80.

La Caze, A. 2011, "The role of basic science in evidence-based medicine," *Biology and Philosophy*, vol. 26, no. 1, pp. 81–98.

Latour, B. 2005, *Reassembling the social: an introduction to actor-network-theory*, Oxford University Press, Oxford; New York.

Latour, B. 1993, *We have never been modern*, Harvard University Press, Cambridge, MA.

Laudan, L. 1977, *Progress and its problems: toward a theory of scientific growth*, University of California Press, Berkeley.

Lazovich, D., Solomon, C.C., Thomas, D.B., Moe, R.E., & White, E. 1999, "Breast conservation therapy in the United States following the 1990 National Institutes of Health Consensus Development Conference on the treatment of patients with early stage invasive breast carcinoma," *Cancer*, vol. 86, no. 4, pp. 628–37.

Leape, L.L., Park, R.E., Kahan, J.P., & Brook, R.H. 1992, "Group judgements of appropriateness: the effect of panel composition," *Quality Assurance in Health Care*, vol. 4, no. 2, pp. 151–9.

Leder, D. 1990, *The absent body*, University of Chicago Press, Chicago.

Lehmann, F., Lacombe, D., Therasse, P., & Eggermont, A.M. 2003, "Integration of translational research in the European Organization for Research and Treatment of Cancer Research (EORTC) clinical trial cooperative group mechanisms," *Journal of Translational Medicine*, vol. 1, no. 1, p. 2.

LeLorier, J., Gregoire, G., Benhaddad, A., Lapierre, J., & Derderian, F. 1997, "Discrepancies between meta-analyses and subsequent large randomized, controlled trials," *New England Journal of Medicine*, vol. 337, no. 8, pp. 536–42.

Leshner, A., Davidson, E., Eastman, P., Grundy, S., Kramer, B., McGowan, J., Penn, A., Strauss, S., & Woolf, S. 1999, *Report of the working group of the advisory committee to the director to review the Office of Medical Applications of Research*, <http://www.nih.gov/about/director/060399a.htm> accessed October 30, 2014.

Lexchin, J., Bero, L.A., Djulbegovic, B., & Clark, O. 2003, "Pharmaceutical industry sponsorship and research outcome and quality: systematic review," *BMJ (Clinical Research Ed.)*, vol. 326, no. 7400, pp. 1167–70.

Liang, M.H. 2003, "Translational research: getting the word and the meaning right," *Arthritis and Rheumatism*, vol. 49, no. 5, pp. 720–1.

Lindee, S. & Mueller, R. 2011, "Is cystic fibrosis genetic medicine's canary?" *Perspectives in Biology and Medicine*, vol. 54, no. 3, pp. 316–31.

Lindemann, H. 1997, *Stories and their limits: narrative approaches to bioethics*, Routledge, New York.

Lindley, D.V. 1982, "The role of randomization in inference," *PSA: Proceedings of the Biennial Meeting of the Philosophy of Science Association*, vol. 1982, no. 2: Symposia and Invited Papers, pp. 431–46.

List, C. 2005, "Group knowledge and group rationality: a judgment aggregation perspective," *Episteme*, vol. 2, no. 1, pp. 25–38.

Longino, H.E. 2002, *The fate of knowledge*, Princeton University Press, Princeton, NJ.

Longino, H.E. 1990, *Science as social knowledge: values and objectivity in scientific inquiry*, Princeton University Press, Princeton, NJ.

Lowe, C.U. 1980a, "The Consensus Development Programme: technology assessment at the National Institute of Health," *BMJ*, vol. 280, no. 6231, pp. 1583–4.

Lowe, C.U. 1980b, "The Consensus Development Program at NIH," *The Lancet*, vol. 315, no. 8179, p. 1184.

Machamer, P., Darden, L., & Craver, C. 2000, "Thinking about mechanisms," *Philosophy of Science*, vol. 67, no. 1, pp. 1–25.

MacIntyre, A.C. 1981, *After virtue: a study in moral theory*, University of Notre Dame Press, Notre Dame, IN.

MacLehose, R.R., Reeves, B.C., Harvey, I.M., Sheldon, T.A., Russell, I.T., & Black, A.M. 2000, "A systematic review of comparisons of effect sizes derived from randomised and non-randomised studies," *Health Technology Assessment*, vol. 4, no. 34, pp. 1–154.

Maienschein, J., Sunderland, M., Ankeny, R.A., & Robert, J.S. 2008, "The ethos and ethics of translational research," *The American Journal of Bioethics*, vol. 8, no. 3, pp. 43–51.

Mankoff, S.P., Brander, C., Ferrone, S., & Marincola, F.M. 2004, "Lost in translation: obstacles to translational medicine," *Journal of Translational Medicine*, vol. 2, no. 1, p. 14.

Mansbridge, J. 1990, "Feminism and democracy," *The American Prospect*, vol. 1, pp. 126–39.

Marincola, F.M. 2007, "In support of descriptive studies; relevance to translational research," *Journal of Translational Medicine*, vol. 5, p. 21.

Marincola, F.M. 2003, "Translational medicine: a two-way road," *Journal of Translational Medicine*, vol. 1, no. 1, p. 1.

Marks, H.M. 1997, *The progress of experiment: science and therapeutic reform in the United States, 1900–1990*, Cambridge University Press, Cambridge; New York.

McClimans, L. 2010, "A theoretical framework for patient-reported outcome measures," *Theoretical Medicine and Bioethics*, vol. 31, no. 3, pp. 225–40.

McClimans, L.M. & Browne, J. 2011, "Choosing a patient-reported outcome measure," *Theoretical Medicine and Bioethics*, vol. 32, no. 1, pp. 47–60.

Meador, C.K. 2005, *Symptoms of unknown origin: a medical odyssey*, 1st edition, Vanderbilt University Press, Nashville, TN.

Meldrum, M.L. 2000, "A brief history of the randomized controlled trial: from oranges and lemons to the gold standard," *Hematology/Oncology Clinics of North America*, vol. 14, no. 4, pp. 745–60.

Meldrum, M.L. 1998, "'A calculated risk': the Salk polio vaccine field trials of 1954," *BMJ*, vol. 317, pp. 1233–6.

Meza, J.P. & Passerman, D.S. 2011, *Integrating narrative medicine and evidence-based medicine: the everyday social practice of healing*, Radcliffe Publishing, London; New York.

Mintzberg, H. 1981, "Organizational design, fashion or fit?" *Harvard Business Review*, vol. 59, no. 1, pp. 103–16.

Montgomery, K. 2006, *How doctors think: clinical judgment and the practice of medicine*, Oxford University Press, Oxford; New York.

Montgomery, K. 1991, *Doctors' stories: the narrative structure of medical knowledge*, Princeton University Press, Princeton, NJ.

Moreno, J.D. 1995, *Deciding together: bioethics and moral consensus*, Oxford University Press, New York.

Moss, S.M., Cuckle, H., Evans, A., Johns, L., Waller, M., Bobrow, L., & Trial Management Group 2006, "Effect of mammographic screening from age 40 years on breast cancer mortality at 10 years' follow-up: a randomised controlled trial," *Lancet*, vol. 368, no. 9552, pp. 2053–60.

Mullan, F. & Jacoby, I. 1985, "The town meeting for technology: the maturation of consensus conferences," *JAMA: The Journal of the American Medical Association*, vol. 254, no. 8, pp. 1068–72.

Nagel, T. 1974, "What is it like to be a bat?" *Philosophical Review*, vol. 83, no. 4, pp. 435–50.

Nelkin, D. & Lindee, M.S. 1995, *The DNA mystique: the gene as a cultural icon*, Freeman, New York.

Nelson, H.L. 2001, *Damaged identities, narrative repair*, Cornell University Press, Ithaca, NY.

Nelson, L.H. 1997, "Empiricism without dogmas," in *Feminism, science and the philosophy of science*, eds. L.H. Nelson & J. Nelson, Kluwer, Dordrecht, pp. 95–119.

Nersessian, N.J. 2008, *Creating scientific concepts*, MIT Press, Cambridge, MA.

Nussenblatt, R.B., Marincola, F.M., & Schechter, A.N. 2010, "Translational medicine: doing it backwards," *Journal of Translational Medicine*, vol. 8, p. 12.

Oliver, M.F. 1985, "Consensus or nonconsensus conferences on coronary heart disease," *The Lancet*, vol. 325, no. 8437, pp. 1087–9.

Oreskes, N. 2004, "Beyond the ivory tower: the scientific consensus on climate change," *Science*, vol. 306, no. 5702, p. 1686.

Pardridge, W.M. 2003, "Translational science: what is it and why is it so important?" *Drug Discovery Today*, vol. 8, no. 18, pp. 813–15.

Pawson, R. & Tilley, N. 1997, *Realistic evaluation*, Sage, London; Thousand Oaks, CA.

Pellegrino, E.D., Engelhardt, H.T., & Jotterand, F. 2008, *The philosophy of medicine reborn: a Pellegrino reader*, University of Notre Dame Press, Notre Dame, IN.

Perry, S. 1988, "Consensus development: an historical note," *International Journal of Technology Assessment in Health Care*, vol. 4, no. 4, pp. 481–4.

Perry, S. 1987, "The NIH consensus development program: a decade later," *New England Journal of Medicine*, vol. 317, no. 8, pp. 485–8.

Perry, S. & Kalberer, J.T., Jr. 1980, "The NIH consensus-development program and the assessment of health-care technologies: the first two years," *New England Journal of Medicine*, vol. 303, no. 3, pp. 169–72.

Pickering, A. 1995, *The mangle of practice: time, agency, and science*, University of Chicago Press, Chicago.

Pickering, A. 1984, *Constructing quarks: a sociological history of particle physics*, University of Chicago Press, Chicago.

Pizzo, P. 2002, "Letter from the Dean," *Stanford Medicine Magazine*, Fall 2002, available at <http://sm.stanford.edu/archive/stanmed/2002fall/letter.html> accessed October 30, 2014.

Polanyi, M. 1958, *Personal knowledge: towards a post-critical philosophy*, University of Chicago Press, Chicago.

Portnoy, B., Miller, J., Brown-Huamani, K., & DeVoto, E. 2007, "Impact of the National Institutes of Health Consensus Development Program on stimulating National Institutes of Health-funded research, 1998 to 2001," *International Journal of Technology Assessment in Health Care*, vol. 23, no. 3, pp. 343–8.

Prendergast, A., Walker, A.S., Mulenga, V., Chintu, C., & Gibb, D.M. 2011, "Improved growth and anemia in HIV-infected African children taking cotri-moxazole prophylaxis," *Clinical Infectious Diseases: An Official Publication of the Infectious Diseases Society of America*, vol. 52, no. 7, pp. 953–6.

Quanstrum, K.H. & Hayward, R.A. 2010, "Lessons from the mammography wars," *New England Journal of Medicine*, vol. 363, no. 11, pp. 1076–9.

Rawlins, M. 2008, "De testimonio: on the evidence for decisions about the use of therapeutic interventions," *Lancet*, vol. 372, no. 9656, pp. 2152–61.

Rawls, J. 1971, *A theory of justice*, Belknap Press of Harvard University Press, Cambridge, MA.

Reichenbach, H. 1938, *Experience and prediction: an analysis of the foundations and the structure of knowledge*, University of Chicago Press, Chicago.

Rennie, D. 1981, "Consensus statements," *New England Journal of Medicine*, vol. 304, no. 11, pp. 665–6.

Riesenberg, D. 1987, "Consensus conferences," *JAMA: The Journal of the American Medical Association*, vol. 258, no. 19, pp. 2738.

Riesman, D. 1931, "The art of medicine," *Science*, vol. 74, no. 1920, pp. 373–80.

Rimmon-Kenan, S. 2006, "What can narrative theory learn from illness narratives?" *Literature and Medicine*, vol. 25, no. 2, pp. 241–54.

Rizzolatti, G., Fogassi, L., & Gallese, V. 2006, "Mirrors in the mind," *Scientific American*, vol. 295, no. 5, pp. 54–61.

Rogers, E.M., Larsen, J.K., & Lowe, C.U. 1982, "The consensus development process for medical technologies: a cross-cultural comparison of Sweden and the United States," *JAMA: The Journal of the American Medical Association*, vol. 248, no. 15, pp. 1880–2.

Rogers, W. 2004, "Evidence-based medicine and women: do the principles and practice of EBM further women's health?" *Bioethics*, vol. 18, no. 1, pp. 50–71.

Rollin, B. 1976, *First, you cry*, 1st edition, Lippincott, Philadelphia.

Rosenberg, C.E. 2007, *Our present complaint: American medicine, then and now*, Johns Hopkins University Press, Baltimore.

Sackett, D.L., Rosenberg, W.M., Gray, J.A., Haynes, R.B., & Richardson, W.S. 1996, "Evidence based medicine: what it is and what it isn't," *BMJ (Clinical Research Ed.)*, vol. 312, no. 7023, pp. 71–2.

Sacks, O.W. 1985, *The man who mistook his wife for a hat and other clinical tales*, Summit Books, New York.

Schechtman, M. 1996, *The constitution of selves*, Cornell University Press, Ithaca, NY.

Schersten, T. 1990, "Topic and scope of consensus development conferences: criteria and approach for selection of topics and properties for assessment," in *Improving consensus development for health technology assessment: an international perspective*, eds. C. Goodman & S. Baratz, National Academy Press, Washington, DC, pp. 18–22.

Schully, S.D., Benedicto, C.B., Gillanders, E.M., Wang, S.S., & Khoury, M.J. 2011, "Translational research in cancer genetics: the road less traveled," *Public Health Genomics*, vol. 14, no. 1, pp. 1–8.

Schunemann, H.J., Oxman, A.D., Brozek, J., Glasziou, P., Jaeschke, R., Vist, G.E., Williams, J.W., Jr., Kunz, R., Craig, J., Montori, V.M., Bossuyt, P., Guyatt, G.H., & GRADE Working Group 2008, "Grading quality of evidence and strength of recommendations for diagnostic tests and strategies," *BMJ (Clinical Research Ed.)*, vol. 336, no. 7653, pp. 1106–10.

Scott, E.A. & Black, N. 1991, "Appropriateness of cholecystectomy in the United Kingdom: a consensus panel approach," *Gut*, vol. 32, pp. 1066–70.

Segal, J. 2007, "Breast cancer narratives as public rhetoric: genre itself and the maintenance of ignorance," *Linguistics and the Human Sciences*, vol. 3, no. 1, pp. 3–23.

Sehon, S.R. & Stanley, D.E. 2003, "A philosophical analysis of the evidence-based medicine debate," *BMC Health Services Research*, vol. 3, no. 14.

Shapin, S. 2008, *The scientific life: a moral history of a late modern vocation*, University of Chicago Press, Chicago.

Shapin, S. 1996, *The scientific revolution*, University of Chicago Press, Chicago.

Shapin, S. 1994, *A social history of truth: civility and science in seventeenth-century England*, University of Chicago Press, Chicago.

Shrier, I., Boivin, J.F., Steele, R.J., Platt, R.W., Furlan, A., Kakuma, R., Brophy, J., & Rossignol, M. 2007, "Should meta-analyses of interventions include observational studies in addition to randomized controlled trials? A critical examination of underlying principles," *American Journal of Epidemiology*, vol. 166, no. 10, pp. 1203–9.

Sismondo, S. 2008, "How pharmaceutical industry funding affects trial outcomes: causal structures and responses," *Social Science & Medicine*, vol. 66, no. 9, pp. 1909–14.

Smith, G.C. & Pell, J.P. 2003, "Parachute use to prevent death and major trauma related to gravitational challenge: systematic review of randomised controlled trials," *BMJ (Clinical Research Ed.)*, vol. 327, no. 7429, pp. 1459–61.

Sniderman, A.D. 1999, "Clinical trials, consensus conferences, and clinical practice," *Lancet*, vol. 354, no. 9175, pp. 327–30.

Snow, C.P. 1959, *The two cultures and the scientific revolution*, Cambridge University Press, New York.

Solomon, M. 2014, "Evidence-based medicine and mechanistic reasoning in the case of cystic fibrosis," in *Logic, methodology and philosophy of science: proceedings of the fourteenth international congress (Nancy)*, eds. P. Schroeder-Heister, W. Hidges, G. Heinzmann, & P.E. Bour, College Publications, London, 2014, page numbers not yet available.

Solomon, M. 2012, " 'A troubled area': understanding the controversy over screening mammography for women aged 40–49," in *Epistemology: contexts, values, disagreement. Proceedings of the 34th International Ludwig Wittgenstein Symposium*, eds. C. Jager & W. Loffler, Ontos Verlag, Heusenstamm, pp. 271–84.

Solomon, M. 2011a, "Group judgment and the medical consensus conference," in *Elsevier handbook on philosophy of medicine*, ed. F. Gifford, Elsevier, Amsterdam, pp. 239–54.

Solomon, M. 2011b, "Just a paradigm: evidence based medicine meets philosophy of science," *European Journal of Philosophy of Science*, vol. 1, no. 3, pp. 451–66.

Solomon, M. 2009, "Standpoint and creativity," *Hypatia*, vol. 24, no. 4, pp. 226–37.

Solomon, M. 2008, "Epistemological reflections on the art of medicine and narrative medicine," *Perspectives in Biology and Medicine*, vol. 51, no. 3, pp. 406–17.

Solomon, M. 2007, "The social epistemology of NIH consensus conferences," in *Establishing medical reality: methodological and metaphysical issues in philosophy of medicine*, eds. H. Kincaid & J. McKitrick, Springer, Dordrecht, pp. 167–77.

Solomon, M. 2006, "Groupthink versus the wisdom of crowds: the social epistemology of deliberation and dissent," *The Southern Journal of Philosophy*, vol. 44, Supplement, pp. 28–42.

Solomon, M. 2001, *Social empiricism*, MIT Press, Cambridge, MA.

Spencer, A.P. & Wingate, S. 2005, "Cardiovascular drug therapy in women," *The Journal of Cardiovascular Nursing*, vol. 20, no. 6, pp. 408–17; quiz, pp. 418–19.

Stegenga, J. 2011, "Is meta-analysis the platinum standard of evidence?" *Studies in History and Philosophy of Biological and Biomedical Sciences*, vol. 42, pp. 497–507.

Straus, S.E. & McAlister, F.A. 2000, "Evidence-based medicine: a commentary on common criticisms," *Canadian Medical Association Journal*, vol. 163, no. 7, pp. 837–41.

Sulmasy, D.P. 2002, "A biopsychosocial-spiritual model for the care of patients at the end of life," *The Gerontologist*, vol. 42, no. 3, pp. 24–33.

Sunstein, C.R. 2003, *Why societies need dissent*, Harvard University Press, Cambridge, MA; London.

Surowiecki, J. 2004, *The wisdom of crowds: why the many are smarter than the few and how collective wisdom shapes business, economies, societies, and nations*, Doubleday, New York; London.

Svenaeus, F. 2000, *The hermeneutics of medicine and the phenomenology of health: steps towards a philosophy of medical practice*, Kluwer Academic Publishers, Dordrecht; Boston.

Thagard, P. 1993, "Societies of minds: science as distributed computing," *Studies in the History and Philosophy of Science*, vol. 24, no. 1, pp. 49–67.

Thamer, M., Ray, N.F., Henderson, S.C., Rinehart, C.S., Sherman, C.R., & Ferguson, J.H. 1998, "Influence of the NIH Consensus Conference on Helicobacter pylori on physician prescribing among a Medicaid population," *Medical Care*, vol. 36, no. 5, pp. 646–60.

Timmermans, S. & Berg, M. 2003, *The gold standard: the challenge of evidence-based medicine and standardization in health care*, Temple University Press, Philadelphia, PA.

Tonelli, M.R. 1999, "In defense of expert opinion," *Academic Medicine*, vol. 74, no. 11, pp. 1187–92.

Tonelli, M.R. 1998, "The philosophical limits of evidence-based medicine," *Academic Medicine*, vol. 73, no. 12, pp. 1234–40.

Toombs, S.K. 2001, *Handbook of phenomenology and medicine*, Kluwer Academic, Dordrecht; Boston.

Toombs, S.K. 1992, *The meaning of illness: a phenomenological account of the different perspectives of physician and patient*, Kluwer Academic Publishers, Dordrecht; Boston.

Toulmin, S.E. 1958, *The uses of argument*, Cambridge University Press, Cambridge.

Upshur, R.E.G. 2003, "Are all evidence-based practices alike? Problems in the ranking of evidence," *Canadian Medical Association Journal*, vol. 169, no. 7, pp. 672–3.

Vandenbroucke, J.P. 2011, "Why do the results of randomised and observational studies differ?" *BMJ (Clinical Research Ed.)*, vol. 343, pp. d7020.

Vandenbroucke, J.P. 2009, "The HRT controversy: observational studies and RCTs fall in line," *The Lancet*, vol. 373, no. 9671, pp. 1233–5.

Wang, X. & Marincola, F.M. 2012, "A decade plus of translation: what do we understand?" *Clinical and Translational Medicine*, vol. 1, no. 3, open access article available at <http://www.clintransmed.com/content/1/1/3> accessed October 30, 2014.

Wehling, M. 2011, "Drug development in the light of translational science: shine or shade?" *Drug Discovery Today*, vol. 16, no. 23–4, pp. 1076–83.

Wehling, M. 2009, "Assessing the translatability of drug projects: what needs to be scored to predict success?" *Nature Reviews Drug Discovery*, vol. 8, no. 7, pp. 541–6.

Wehling, M. 2008, "Translational medicine: science or wishful thinking?" *Journal of Translational Medicine*, vol. 6, article 31, <http://www.translational-medicine.com/content/pdf/1479-5876-6-31.pdf> accessed November 5, 2014.

Wehling, M. 2006a, "Translational medicine: can it really facilitate the transition of research 'from bench to bedside'?" *European Journal of Clinical Pharmacology*, vol. 62, no. 2, pp. 91–5.

Wehling, M. 2006b, "Translational science in medicine: implications for the pharmaceutical industry," *International Journal of Pharmaceutical Medicine*, vol. 20, no. 5, pp. 303–10.

Weissman, G. 2005, "Roadmaps, translational research and childish curiosity," *The FASEB Journal*, vol. 13, November, pp. 1761–2.

Welch, H.G. & Frankel, B.A. 2011, "Likelihood that a woman with screen-detected breast cancer has had her 'life saved' by that screening," *Archives of Internal Medicine*, vol. 171, no. 22, pp. 2043–6.

Wells, J. 1998, "Mammography and the politics of randomised controlled trials," *BMJ (Clinical Research Ed.)*, vol. 317, no. 7167, pp. 1224–9.

Wendler, A. & Wehling, M. 2012, "Translatability scoring in drug development: eight case studies," *Journal of Translational Medicine*, vol. 10, p. 39.

Wendler, A. & Wehling, M. 2010, "The translatability of animal models for clinical development: biomarkers and disease models," *Current Opinion in Pharmacology*, vol. 10, no. 5, pp. 601–6.

Wexler, N. 1992, "Clairvoyance and caution: repercussions from the Human Genome Project," in *The code of codes: scientific and social issues in the Human Genome Project*, eds. D. Kevles & L. Hood, pp. 211–43.

White, M. & Epston, D. 1990, *Narrative means to therapeutic ends*, Norton, New York.

Woolf, S.H. 2008, "The meaning of translational research and why it matters," *JAMA: The Journal of the American Medical Association*, vol. 299, no. 2, pp. 211–13.

Worrall, J. 2007a, "Evidence in medicine and evidence-based medicine," *Philosophy Compass*, vol. 2, no. 6, pp. 981–1022.

Worrall, J. 2007b, "Why there's no cause to randomize," *The British Journal for the Philosophy of Science*, vol. 58, no. 3, pp. 451–88.

Worrall, J. 2002, "What evidence in evidence-based medicine?" *Philosophy of Science*, vol. 69, no. S3, pp. 316–30.

Worthington, R. 2011, *Letter to the science, innovation and further education committee of the Danish parliament*, <http://www.loka.org/Documents/ US%20Scientists%20Letter%20to%20Folketinget%20Dec%202011.pdf> accessed November 4, 2014.

Wortman, P.M., Vinokur, A., & Sechrest, L. 1988, "Do consensus conferences work? A process evaluation of the NIH Consensus Development Program," *Journal of Health Politics, Policy and Law*, vol. 13, no. 3, pp. 469–98.

Wylie, A. 2002, *Thinking from things: essays in the philosophy of archaeology*, University of California Press, Berkeley.

Wylie, A. 2000, "Rethinking unity as a 'working hypothesis' for philosophy of science: how archaeologists exploit the disunities of science," *Perspectives on Science*, vol. 7, no. 3, pp. 293–317.

Yank, V., Rennie, D., & Bero, L.A. 2007, "Financial ties and concordance between results and conclusions in meta-analyses: retrospective cohort study," *BMJ (Clinical Research Ed.)*, vol. 335, no. 7631, pp. 1202–5.

Young, I.M. 2001, "Activist challenges to deliberative democracy," *Political Theory*, vol. 29, no. 5, pp. 670–90.

Zaner, R.M. 2004, *Conversations on the edge: narratives of ethics and illness*, Georgetown University Press, Washington, DC.

Zerhouni, E. 2003, "Medicine: the NIH Roadmap," *Science*, vol. 302, no. 5642, pp. 63–72.

# Index

Printed and bound by CPI Group (UK) Ltd, Croydon, CR0 4YY